John Budge

The practical miners' guide

Comprising a set of trigonometrical tables

John Budge

The practical miners' guide
Comprising a set of trigonometrical tables

ISBN/EAN: 9783337203092

Printed in Europe, USA, Canada, Australia, Japan

Cover: Foto ©berggeist007 / pixelio.de

More available books at **www.hansebooks.com**

THE

PRACTICAL MINER'S GUIDE:

COMPRISING A SET OF TRIGONOMETRICAL TABLES ADAPTED
TO ALL PURPOSES OF OBLIQUE OR DIAGONAL,
VERTICAL, HORIZONTAL, AND TRAVERSE

DIALLING,

WITH THEIR APPLICATION TO THE DIAL EXERCISE OF SHAFTS, ADITS, DRIFTS, LODES,
SLIDES, LEVELLING, INACCESSIBLE DISTANCES, HEIGHTS, ETC.

Also a Treatise on the Art and Practice of

ASSAYING SILVER, COPPER, LEAD, AND TIN,

WITH TABLES WHICH EXHIBIT AT ONE VIEW THE VALUE OF ASSAYED ORES, RULES
FOR CALCULATING THE POWER OF STEAM AND WATER ENGINES, A DISCOURSE ON
THE QUALITY, MANUFACTURE, AND CHOICE OF CORDAGE FOR MINE SERVICE;

Together with a Collection of Essential

TABLES, RULES, AND ILLUSTRATIONS

EXCLUSIVELY APPLICABLE TO MINING BUSINESS.

The whole introduced and exemplified in the most plain and practicable manner.

BY J. BUDGE.

NEW EDITION.

LONDON:
LONGMANS, GREEN, AND CO.
1866.

TO

BENJAMIN TUCKER, ESQ.

SURVEYOR-GENERAL OF THE DUCHY OF CORNWALL,

AND

JOHN TAYLOR, ESQ.

DIRECTOR OF BRITISH AND FOREIGN MINING INSTITUTIONS.

---- ◆ ---

GENTLEMEN,

THE propriety of composing an address
under this head did not, until of late, occur to my
imagination, and with much trepidity and circumspection do I now attempt the difficult task. My
apprehensions arise from a consciousness of the extreme delicacy of the undertaking, and the danger
I have to encounter, arising from the probability of
a misconstruction being put on the sentiments I
may advance; nevertheless I feel an enlivening
hope, springing up as I proceed, that I shall not be
unhappy enough to make an insertion of a single
sentence but what you will either approve or pal-
liate, and, this point secured, I shall be compara-
tively indifferent of all other consequences.

The succeeding remarks will prove explanatory
of my motives, and I trust will be deemed a satis-

factory apology for my adventuring into the dedi-
catory labyrinth, from whence but few escape
without considerable loss in the estimation of the
most discerning part of their readers.

It is a prevalent opinion that gentlemen sometimes allow authors to make an ostentatious display
of their influential names in the front of their
works, while at the same time they possess not the
slightest knowledge of the composition, or of ever
having seen a line thereof, until a finished copy
has been formally presented them, adorned with
their illustrious names and the writer's adulatory
compliments; but further, it is notorious that com-
posers not unfrequently take the unwarrantable
franchise of introducing the honourable names of
eminent persons for this purpose without ever
obtaining or even soliciting their permission for so
doing.

These considerations have induced me, with great
diffidence, and contrary to my preconception of
the matter, to frame this respectful address; for,
after mature deliberation, I have been led to conclude that it would be doing an act of injustice to
you, to myself, and to those who may be interested
in the contents of this volume, were I to forbear
giving publicity to the fact, that it was not until
you had investigated the most essential parts of the
manuscript, and felt assured that it was calculated
to answer the end designed, and thereby prove

advantageous to the mining interest, that you con-
sented to honour it with your countenance, and
allow your respected names to be introduced in a
patronal capacity.

Without fear of the imputation of flattery, I
may with confidence remark, that your great experience, knowledge, celebrity, and interest in mining
affairs, are calculated to inspire a reasonable hope
that, through your support and precedent, success
will attend my humble efforts, and I shall ultimately obtain a fair pecuniary remuneration for my
labours; but, necessitous as I am, the animating
thought of my exertions becoming instrumental in
yielding an important benefit to a considerable and
most useful class of my fellow-men, is by far the
greatest cause of my exultation.

It now only remains for me to acknowledge the
obligations I feel, not for mere nominal patronage,
but for the actual assistance I have received at
your hands, whereby I have been enabled to over-
come the formidable obstacles which presented
themselves against the accomplishment of this difficult and embarrassing enterprise. And at the
same time I would not dare neglect to return my
unfeigned thanks, in the most undisguised and so-
lemn manner, to that supreme and gracious Power
whose benign and omnipotent influence has been
manifestly exercised in my behalf throughout the
arduous undertaking.

Possibly this digression may be condemned as unseasonable, unsuitable, and *unfashionable*, by those who fancy themselves too wise and independent to allow it to be obligatory upon them openly to acknowledge the aid of Divine Providence in all their successful engagements; but desirous as I am of giving universal satisfaction, far rather will I endure the scorn and sarcasms of such indi v duals, than avoid their contumely at the expense of an approving conscience, and the approbation of the truly wise and honourable.

In conclusion, I sincerely beg excuse for the almost unprecedented latitude I have unintentionally taken for this subject, and allow me most cordially to congratulate you on the high degree of respectability as well as the national utility and importance the mining system of Great Britain has recently acquired, and to which, it is generally acknowledged, your example and instrumentality have materially contributed.

That you may live long and prosperously, and that uninterrupted success may attend your every endeavour to promote your own and the public welfare, is the sincere and hearty desire of,

Gentlemen,
Your most grateful,
most obedient,
and very humble servant,
THE AUTHOR.

PREFACE.

THROUGHOUT this volume the benefit of the Practical Miner has been constantly kept in view; and although the Tables may be successfully used on other occasions, yet we have professedly the mining interest solely for our object.

It is hoped that the curious examiner will find no flagrant violation of the rules of composition or mathematical order; but to any who may be disposed to search for defects we beg to state, that the work has been prosecuted and accomplished amidst numerous disadvantages, among which, the frequent interruptions arising from the almost incessant duties of a mining avocation have not been the least perplexing. But, notwithstanding all this, we feel it incumbent on us to state, in defence of the calculations, that, so far as Hutton's celebrated Logarithms, compared with those of other eminent authors, can be relied on as a foundation, together with the utmost care in, and repetition of, every single operation, supported by numerous subsequent proofs, both theoretical and practical, we are warranted in alleging our conviction, that the Tables may be received and applied to the most difficult and important operations in dialling, with the utmost confidence.

We now take a transient, but submissive leave of
the Mine Agent, and respectfully address ourselves
to persons in the capacity of Schoolmasters in mining districts.

To that respectable class of men this work is also
recommended, as a fit subject for the study of those
youths committed to their instruction, who appear
designed for a mining occupation; and having ourselves had some years' experience in public teaching,
we presume to declare our persuasion, that it is
capable of being rendered extremely useful, by the
prudent tuition of skilful preceptors. And why,
we would inquire, should not the theory of the
essential art of mining be taught in the seminaries
of England, as well as in Mexico? (where, we
understand, the colleges or schools of mines are
among the most noble establishments of the em-
pire); but especially in Cornwall, where the principal support of the country depends on the
prosperity of the mines, and where much of that
prosperity depends on the judicious management
of those persons to whom the superintendency thereof is committed.

In conclusion, we beg to remark, that we have
generally used such terms as are common, and in
some cases almost peculiar, to mining business;
believing that this phraseology will render the work
far more intelligible and acceptable to the majority
of those for whose use it is designed, than if we
had confined ourselves to a precise mathematical
mode of expression.

CONTENTS.

PART I.

CONTENTS. xiii

PAGE

Method of extracting Silver from Copper Ore . . . 110

Smelting Silver Ore 110

„ Lead and Silver Ore 110

Silver Assay Table 111

Method of computing the Value of Lead and Silver Ore . 115

„ „ „ Copper Ore . . 116

Rule for discovering the Power of Steam Engines . . 117

„ „ „ Water Engines . . 119

Table of the Square Inches in a Cylinder 123

Table of the Weight of Water in a Pump, &c. . . . 125

Treatise on Cordage 126

Table showing the Number of Threads in a Rope . . 135

„ „ Length of Rope to an Hundred-weight . 136

„ „ Weight of Ropes 100 Fathoms long . 137

Observations on Capstan Machinery 139

SUPPLEMENT, OR THIRD PART.

Introduction 145

Traverse Dialling 149

Table for converting Angles into Bearings . . . 165

Plans and Sections of Mines 190

Miscellaneous 203

Geology 206

EXPLANATION OF TERMS 217

EXPLANATION OF THE DIAGRAMS.

——✦——

TABLE I. Page 66.

In this scheme the hypothenuse is made radius, consequently the other sides are the sine and cosine of the included angle.

Corollary. Suppose one end of the line A B to remain at A while the other end B is moved round from *e* to *f*, then it is evident that the base C B will continue to increase, and the perpendicular B D to decrease, until the whole quadrant has been swept off.

At 45°, or the middle of the quadrant, the base and perpendicular are equal, and from that point to 90° the base will increase in the same ratio as the perpendicular decreased from 1° to 45°; hence the propriety of the arrangement of this table in counting the degrees backward from 45 to 90.

TABLE II. Page 74.

Here the perpendicular is made radius; therefore the hypothenuse A C will be the secant, and the base B C the tangent, of the angle A. On this principle it is clear that as the angle increases, the base and hypothenuse will continue (throughout the whole quadrant) to increase in proportion.

TABLE III. Page 82.

In this diagram the base is made radius, therefore by mathematical demonstration the perpendicular A C is the co-tangent, and the hypothenuse B C the co-secant, of the angle C; and here it will be plain, that as the angle C is increased, the hypothenuse and perpendicular will, proportionably, be diminished.

+ *plus*, or *more*, the sign of addition ; signifying that the numbers or quantities between which it is placed are to be added together.

— *minus*, or *less*, the sign of subtraction ; denoting that the less of the two quantities between which it is placed is to be taken from the greater.

× *into*, the sign of multiplication ; signifying that the quantities between which it is placed are to be multiplied together.

÷ *by*, the sign of division ; signifying that the former of the two quantities between which it is placed is to be divided by the latter.

: *as*, or *to*, :: *so is*, the sign of an equality of ratios ; denoting that the quantities between which they are placed are proportional to each other.

Thus, 2 : 3 :: 4 : 6, denotes that 2 is to 3 as 4 is to 6.

= *equal to*, the sign of equality ; signifying that the quantities between which it is placed are equal to each other.

Thus, 6 + 4 = 10, shows that 6 added to 4 is equal to 10.

∠ *Angle.*

° *Degrees.*

′ *Minutes.*

A given line is represented by a stroke or dash (ı), as the base A B in the triangle on page 26, and a required line by a cipher (○), as in the legs of the same triangle.

ABBREVIATIONS.

Deg.	Degrees.	E.	East.
Min.	Minutes.	W.	West.
Fath.	Fathoms.	Prob.	Problem.
Ft.	Feet.	Ex.	Example.
In.	Inches.	Ansr.	Answer.
Hyp.	Hypothenuse.	Dia.	Diameter.
Perp.	Perpendicular.	Cwt.	Hundred-weight.
Comp.	Complement.	Qr.	Quarter.
Tab.	Tabular.	Lbs.	Pounds.
Dec.	Decimals.	Oz.	Ounces.
N.	North.	Dwt.	Pennyweight.
S.	South.	Grs.	Grains.

PRACTICAL MINER'S GUIDE.

INTRODUCTION.

It is an acknowledged fact, that dialling, in all its varieties, is the most difficult and momentous part of the duty of practical mine agents : to assist them in that important operation, is the chief design in giving publicity to these Tables.

Notwithstanding the great improvements which of late years have been made in the art of dialling, the most intelligent miners universally admit, that the practice is still very imperfect; nay, so far are they from any determinate and general system, that two persons can scarcely be found who precisely adopt the same method; consequently some plain scheme, founded on pure mathematical principles, is a great mining desideratum.

Aware of the opposition which is so apt to arise against all attempts at innovation of an old and established habit (which, however faulty in itself, custom may have stamped with an imaginary perfection), it may be necessary to make a few observations in support of this work, and endeavour to prove its advantages over all the preceding

B

modes of performing a dialling operation in every respect.

Not many years ago the customary way of ascertaining the perpendicular and horizontal lines corresponding with a diagonal shaft, was by the very uncertain, expensive, and tardy practice of dropping a plumb-line from the back to the bottom, there fixing a sollar or platform, and repeating the process from the brace to the foot of the shaft; this usage is largely explained in ʻ Pryce's Treatise on Mines and Mining' (a celebrated work, published by subscription, about the year 1776), and therein described as the only system then known.

It is true this most objectionable measure is now exploded, but not without great reluctance by many of its old practitioners, who were long before they could be prevailed on to abandon it, notwithstanding its glaring inconveniences, accompanied with the loss of time, waste of property, and hazard of their lives.

By inserting this defectibility of our predecessors, we have no other design than of cautioning our mining countrymen to guard against the too prevalent propensity of rejecting any new system, merely because it is *new*, or its utility not discerned at first sight; and to induce them to give the subject an impartial investigation, before they pass a conclusive judgment thereon.

The use of mathematical instruments is now partially known in the mining world; and, certainly, those agents who are well acquainted therewith, possess a decided advantage over others who are

not ; for, doubtless, this science has the pre-
eminence, in a high degree, over every other
method heretofore employed in dialling. But
without intending to undervalue instrumental
operation, we appeal to the experience of our
scientific readers for support, in avouching that the
process is ever liable to errors of considerable
extent, and which are prone to slide in unaccount-
ably : but it is a palpable fact, that in pointing or
sweeping off the angles, an almost imperceptible
deviation will create a serious departure from truth ;
and even in the course of bisecting, trisecting, in-
scribing, describing, and circumscribing ; also in
drawing parallels, raising or demitting perpendiculars, the operation, even with the greatest care,
is exposed to considerable mistakes : and so sensible
are all professional men of this defect, that instrumental operation is never resorted to, or relied on,
in any case where great accuracy is required. But
when we reflect on the laborious duties of the
practical mine agent, and how much these duties
are calculated to disqualify him for performing a
geometrical plan with that delicacy and precision
which the operation so indispensably demands, we
then become established in our opinion of the necessity of a work of this kind, and of its superiority
over every other system hitherto introduced in
dialling.

Should any be yet disposed to advocate the
existing practice, and to contend that it is fully
adequate to the desired purpose, we beg permission
to inquire of such persons, why it is that mistakes

so commonly occur in sinking shafts and driving levels in most of our mines? That irreparable errors do frequently happen, is a truth too notorious for contradiction or dispute, and sometimes even under the superintendence of men whose knowledge, circumspection, and experience no one presumes to call in question; consequently a more convincing proof than this cannot be adduced of the fallibility of the best modern practice, and the necessity for the introduction of a more perfect system.

Should it be inquired wherein the merit of this work is considered to consist, we answer, First— *Accuracy*; and it will be discovered at a glance, that every operation of the principal tables is wrought out to five places of decimals, or the *ten thousandth part of an inch!* consequently we may affirm, without fear of confutation, that, in this property, we outvie every other system.

Secondly—*Plainness.* Of this quality our expert readers will be convinced at first sight, and will need no instruction for enabling them to apply the numbers readily; but we do not hesitate to say, that, by the help of the rules and examples, a common school-boy will find no insurmountable difficulty in solving the most abstruse problems relevant to dialling.

Thirdly—*Despatch.* To this desirable property no other system has an equal claim, or can, with any chance of success, enter into competition, with our method; inasmuch as an answer, in most cases, may be obtained by the tables in less time than is necessary to make a preparation for performing the operation in any other way.

And now, having briefly endeavoured to set forth the work in a true light, we commit it to the judgment of a liberal and discerning people; and should it be instrumental in happily preventing the grievous errors which are so prevalent in mining operations (and which, we are bold to say, must, in the nature of things, continue to take place by the present day practice); or should it only help to relieve the minds of faithful superintendents from that painful anxiety and suspense which never fail to harass them during the progress of any considerable work, whereby a heavy responsibility rests on them for the accuracy of their dialling; or should it in any other way have the happy tendency of promoting the interest of mining, we shall not regret the labour, pain, expense, privation, trouble, and perplexity it has cost us, even though we should never receive any other compensation.

TABLES.

— ✦ —

AFTER so many preliminary observations, it will be necessary to say but little under this head, having already anticipated several things by way of introduction, which properly belong here.

The reader will observe that the work is composed of three distinct tables, for the obvious reason of making each side of the triangle radius; and certainly without such an arrangement it would have been incomplete.

In each case the radius, or given side, is one fathom, being the most convenient and familiar proportion that could have been introduced.

The principal calculations include every quarter, or fifteen minutes of a degree, and extend from 1 to 89 degrees, being sufficiently extensive and minute for mining purposes (the angle of any intermediate division not being distinguished or required); and here, it must be observed, that the divisions are expressed by 15, 30, and 45 minutes, which numbers represent ¼, ½, and ¾ of a degree.

The first and most essential table is that wherein the hypothenuse, or longest side, is made radius, extending nearly throughout the quadrant, and every calculation wrought out to five decimal places of an inch, hereby giving a direct answer, in exact ratio to six feet of the given side, to the ten thousandth part of an inch.

Perhaps there may be a little difficulty at first, with persons unacquainted with mathematical order, in reading the first table. It must be remarked, that from 1° to 45°, or the middle of the quadrant, the degrees and parts are all on the left-hand side descending, the base stands in the adjoining columns, and the perpendicular on the same line to the right ; but beyond that point the degrees will be found on the right-hand side ascending, and then it must be specially noted that the perpendicular and base will have changed their positions, the base now standing on the right hand, and the perpendicular on the left hand side.

In the second table the perpendicular is given and the angles extend to 60°. One valuable mining property of this table is, that it gives at sight the underlay in a fathom of every angle within the range of 60° including the divisions : so that if it is required to know the underlay in a fathom on any degree, or quarter of a degree, between 1 and 60, it will be immediately discovered by an inspection of the base in the column adjoining the given angle in this table.

In the third and last table the base is given, and as the application of this part of the work is not so general as the preceding, the angles have been given in degrees only : nevertheless this table is

.

indispensable on some occasions, especially in levelling or driving adits. It will be found, like the second table, to extend from 1 to 60 degrees.

Having thus briefly stated the nature of the work under each separate head, it only remains for us, after a few general observations, to recommend the learner to the inspection of the following examples; for we believe that one practical operation will do more towards giving him a clear understanding or comprehension of the subject, than a volume written expressly thereon, confined to mere speculative description.

It may be remarked that, in almost every instance, the geometrical construction of the figure is introduced, with the calculation, which will tend to the satisfaction of the practitioner and improvement of the beginner.

In conclusion we would remark, that the same attention must be paid in taking the angle, and measuring the given line, when these tables are used, as if the operation were performed any other way.

It is a common practice in mining to take the angle of underlaying shafts with the cover of the dial and a plumb-line; and in short drafts, with great care, this method may answer well enough: but when any very important work is to be performed, we would strongly recommend the application of a more perfect instrument for ascertaining the angle; for it is well known, that if this part of the process should not be correct, the result of the whole work must be erroneous as a matter of course; and indeed it is next to impossible to distinguish the minutia of an angle, with any tolerable degree of certainty, by the foregoing method. There doubtless are instruments much better adapted to the work, both for speed and accuracy, than the dial; and it is matter of surprise that they have not been more generally introduced in our mines : of these instruments the Theodolite certainly stands unrivalled for taking both horizontal and vertical angles.

It is not our design to enter into controversy on this subject ; those who imagine the sextant or quadrant graduated on the cover of the dial well calculated for the purpose, let them continue to use it ; only we would especially note, that should an error ensue, it ought by all means to be attributed to the real cause, and to that only : for, as in all trigonometrical questions, the angle and side are always given to find the other parts of the triangle, consequently the sum of the one, and length of the other, are presupposed to have been correctly ascertained, previous to the commencement of any other operation.

Finally, for the learner's sake, we observe, that as the tables exhibit only the relative proportions to the radius of one fathom, or six feet, and are wrought out to five places of decimals to an inch, it becomes necessary that every one who would use this work successfully should have some knowledge of decimated arithmetic ; because he will have, in most cases, to multiply for the whole numbers, and take parts for the fraction of the fathom. For example : suppose the given side to be the hypothenuse, measuring 16 fathoms, 3 feet, and 6 inches, he will then have to take out the numbers opposite the given angle in the tables, and multiply them by 16, for the base and perpendicular respectively, then divide half the tabular measure for the 3 feet, and one sixth of the remainder for the 6 inches, and add them together for the sum of the required sides of the triangle. We have therefore introduced the following rules and examples in decimals, which are sufficient to enable any one hitherto unacquainted with this branch of arithmetic to use the tables with the greatest facility.

REDUCTION OF DECIMALS.

RULE.— Multiply the decimal by the number of parts in the next less denomination, and cut off as many places to the right hand as there are places in the given decimals.

What is the value of ·75014 of a fathom ?

$$\begin{array}{r} ·75014 \\ 6 \\ \hline 4·50084 \\ 12 \\ \hline 6·01008 \end{array}$$

    ft.   in.
Answer 4 6·01008

What is the value of ·93862 of a yard ?

$$\begin{array}{r} ·93862 \\ 3 \\ \hline 2·81586 \\ 12 \\ \hline 9·79032 \end{array}$$

    ft.   in.
Answer 2 9·79032

What is the value of ·27734 of a foot ?

$$\begin{array}{r} ·27734 \\ 12 \\ \hline 3·32808 \end{array}$$

   in.
Answer 3·32808

    fath. ft. in.
Reduce 5   4 6·32 to feet, inches, and decimals.
       6
     — in.
Answer 34 6·32

ADDITION OF DECIMALS.

RULE.— Place the numbers so that the decimal points may stand directly under each other, add up as in Simple Addition, and cut off for decimals as many figures to the right as there are decimals in the greatest given number.

EXAMPLE.

What is the sum of 3·72 and 14·7368 and 146·2 and ·728 and 5·034 ?

$$
\begin{array}{r}
3\cdot72 \\
14\cdot7368 \\
146\cdot2 \\
\cdot728 \\
5\cdot034 \\
\hline
170\cdot4188
\end{array}
$$

	ft.	in.		ft.	in.
What is the sum of	2	11·9942	and	1	4·09658 ?
	1	4·09658			
	4	4·09078			

Add together the following measures ; viz.

fath.	ft.	in. dec.
6	4	2·260
0	1	11·47298
19	0	3·087
64	5	9·9746
0	0	2·70643
91	0	5·50101

SUBTRACTION OF DECIMALS.

RULE. — Arrange and cut off the decimals as in Addition.

EXAMPLE.

	fath.	ft.	in.
From	4	2	9·7824
Take	2	4	8·91773
	1	4	0·86467

MULTIPLICATION OF DECIMALS.

RULE. — Multiply as in whole numbers, and cut off as many figures from the product as there are decimals in the multiplier and multiplicand.

EXAMPLE.

	fath.	ft.	in.
Multiply	2	4	7·92486 by 24
			6
	16	3	11·54916
			4 $\qquad$ 6 × 4 = 24
	66	3	10·19664

	fath.	ft.	in.
Multiply	9	3	1·4872 by 37
		.	6
	57	0	8·9232
			6
	342	4	5·5392
	9	3	1·4872
	352	1	7·0264

Here we multiply by 6 twice because 6 times 6 are 36, and add the given number, which makes it equal to 37, or 6 × 6 + 1 = 37.

	ft.	in.
Multiply	14	9·746 by 12
		12
	177	8·952

DIVISION OF DECIMALS.

RULE.—Divide as in whole numbers, and cut off as many figures in the quotient as the decimal places in the dividend exceed those of the divisor.

EXAMPLE.

	fath.	ft.	in.	
Divide	2	4	3·7	by 6

$$6)\overline{2 \quad 4 \quad 3·7}$$
$$0 \quad 2 \quad 8·61$$

fath.	ft.	in.
7)4	2	10·30994
0	3	10·04427

fath.	ft.	in.
8)15	5	0·3316
1	5	10·5414

ALIQUOT PARTS OF A FATHOM.

TABLE.

Parts		Feet	Inches
$\frac{1}{2}$	of a fathom is	3	0
$\frac{1}{3}$	ditto	2	0
$\frac{1}{4}$	ditto	1	6
$\frac{1}{6}$	ditto	1	0
$\frac{1}{8}$	ditto	0	9
$\frac{1}{9}$	ditto	0	8
$\frac{1}{12}$	ditto	0	6
$\frac{1}{16}$	ditto	0	$4\frac{1}{2}$
$\frac{1}{18}$	ditto	0	4
$\frac{1}{24}$	ditto	0	3

It has been observed that the radius in every case is 6 feet or 1 fathom; consequently the number of fathoms in the given side, whether that side be hypothenuse, perpendicular, or base, will be the multiplier of the tabular numbers, and should there be a fraction in the multiplier, the multiplicand must be divided by that fraction agreeably with the rule of practice. The table of aliquot parts of a fathom, in the adjoining page, will be found useful in facilitating this part of the process.

In some of the following examples the product has been obtained in fathoms and parts, but we would recommend the learner to carry on the work in *feet* (except in cases where the answer is required in fathoms), as he will find it more simple and expeditious; we speak of the *multiplicand* or number *multiplied* : the *multiplier* must invariably be fathoms, and should the given side be nominated in feet, it must be divided by 6, to bring it into fathoms, before the operation is begun by the foregoing cases.

It may be further noticed that when any of the given sides in the tables amount to 6 feet, they are expressed in fathoms, &c. : but whenever it may be required to produce the answer in feet, &c., the numbers should be reduced to that measure before they are multiplied, and this can be done by mere inspection; viz.

		fath.	ft.	in.			ft.	in.
Table 2nd,	Base	1	1	8·1560	state	Base	7	8·1560
∠ 52°	Hyp.	1	3	9·6053		Hyp.	9	9·6053

16

PRELIMINARY CHAPTER

TO THE

PRACTICAL DIALLING EXAMPLES.

—⋆—

It must have been matter of regret to every re-
flecting, well-informed, and interested person, that
(previous to the present work) nothing has ever
been published with a design to assist the British
miner in his subterraneous operations ; and while
the press has teemed with publications distinctly
and exclusively adapted to benefit the navigator,
the architect, the sculptor, the surveyor, and even
the mechanic and artisan, not a single effort has
ever been made to extricate the miner from the
disadvantages under which he has ever laboured
(solely for the want of a plain, concise, technical,
and scientific treatise on dialling, accompanied with
appropriate tables), although his profession yields
to none in importance and utility : in fact, it may be
said, in a certain sense, to be the parent of every
art and science in the world ; the use of metallic
substances, in some shape or other, being indispensable in every one of them : nevertheless this
highly essential art has hitherto been totally dis-
regarded by all classes of mathematicians, and
while the famous invention of logarithms has
caused the science of trigonometry to soar to the
very skies, and traverse old ocean's vast and
unfathomable expanse, the unsupported miner has

been left to struggle under the greatest disadvantages, with nearly as little obligation to geometrical science, as his antediluvian progenitors;
and although he has done everything that deep
thought, strong natural understanding, unwearied
perseverance, and inventive genius (unassisted by
trigonometrical demonstration) could possibly accomplish, yet, for the want of mathematical light,
his exertions have been ineffectual and insufficient
to disentangle him from the difficulties with which
he has been encircled ; hence his avocation has, in
general, been replete with toil, anxiety, apprehen-
sion, dissatisfaction, and disappointment.

How far the present work is adapted to answer
the great end in contemplation, must be left for
the judgment of the mining world to decide ; and
we doubt not but the defects (real or imaginary)
which may be considered to exist in the application,
will be passed over and excused by every liberal
man, on the grounds already stated in the preface,
having an unshaken confidence that the fundamen-
tals of the work, comprised in the trigonometrical
tables, will be found plain, true, and unexception-
able.

DEFINITION OF RIGHT-ANGLED TRIANGLES.

In order to use the following tables with due
effect, there is no necessity that the reader should
understand anything of the science of trigonometry, that part of the work having been accomplished already to his hand ; so that, by the help

c

of a few of the common rules of arithmetic, he may obtain, with the greatest ease and certainty, every-thing required to be known in the geometrical part of mining.

Previous to an elucidation of the simple method of working by the tables, it may be satisfactory to introduce the operation by a few preliminary ob-servations and extracts on the nature and pro-perties of right-angled triangles.

Plane trigonometry is the art of measuring the sides and angles of triangles described on a plane surface, or of such triangles as are composed of straight lines.

The theory of triangles is the very foundation of all geometrical knowledge, for all straight-lined figures may be reduced to triangles. The angles of a triangle determine only its relative species, and are measured in degrees, minutes, and seconds ; but the sides determine its absolute magnitude, and may be expressed in fathoms, yards, feet, or any other lineal measure.

THEOREMS.

A right-angled triangle (the only kind generally necessary to be treated of for mining purposes) is that which has one right angle in it ; the longest side, or that opposite to the right angle, is called the hypothenuse, the other two are called the legs or sides, or the base and perpendicular : or, by Euclid's definition, ' In a right-angled triangle, the side opposite to the right angle is called the

HYPOTHENUSE, and of the other sides, that upon which the figure is supposed to stand is called the BASE, and the remaining side the PERPENDICULAR.'

The three angles of every triangle are together equal to two right angles, or 180 degrees.

The greater side of every triangle has the greater angle opposite to it.

The squares of two sides of a triangle are together double of the square of half the base, and of the square of a straight line drawn from the vertex to bisect the base.

The sum of the three angles of every plane triangle being equal to half a circle, or 180 degrees, it therefore follows that if either acute angle, in such triangle, be taken from 90°, the remainder will be the other acute angle, or the complement.

The supplement of any angle is what that angle wants of 180°; hence the supplement of any one angle is always equal to the sum of the other two.

A few other properties of right-angled triangles may be worthy of notice, viz. : when the angle opposite the base is 30°, the hypothenuse is exactly double the length of the base.

When the angles are 45°, the base and perpendicular are equal.

When the angle opposite the base is 60°, the hypothenuse is double the length of the perpendicular.

APPLICATION.

To show how a knowledge of the foregoing theorems may be rendered useful in mining practices, suppose in the triangle A B C, on page 24, the base B A represented a drift or cross-cut, and the side A C a lode, making an angle with the base of 66° 30′; consequently the angle A must be 23° 30′, because it requires that number of dgrees to constitute a right angle, the complement of the angle A, or 180°, the supplement of the triangle A B C.

Again, suppose the angle C of the diagonal shaft C A, page 25, were found to be 39° 30′, then the opposite angle A must contain 50° 30′.

We now approach towards the actual use of the tables, and have succeeded, we hope, in clearing all impediments out of the learner's way, so that he will find no difficulty in readily applying the numbers to dialling operations. We have previously set a few examples of the mere act of taking out the primes, and have studiously endeavoured to render everything as perspicuous and comprehensible as the nature of the work would possibly admit. But should anyone have gone thus far and still find an obscurity hang over him, so that he cannot penetrate into the nature of the subject as he would wish, or as he may have expected, yet let him not be discouraged; this will always be the case with everyone who calculates on fully comprehending anything connected with the mathematics by definition or description only. Let him steadily, attentively, and perseveringly proceed with the examples, and if he is properly interested in the matter, he will soon find the subject open with perspicuity and demonstration on his mind, and convey to him the incontrovertible assurance of the truth of the calculations, as well as the correctness of his own views, ideas, or conceptions of the subject.

TABLE I.—EXAMPLE.

When the angle is 9° and the hypothenuse 1 fathom, what is the length of the other two sides of the triangle respectively?

(page 68.)

	in.		ft.	in.
Answer, Base	11·26328	Perp.	5	11·11356

EXAMPLE.

When the angle is 48° 15', or 48¼ degrees*, and the hypothenuse 1 fathom, what are the lengths of the other sides?

(page 72.)

	ft.	in.		ft.	in.
Answer, Base	4	5·71613	Perp.	3	11·94348

TABLE II.—EXAMPLE.

When the angle is 35° 45', or 35¾ degrees, and the perpendicular 1 fathom, what is the length of the hypothenuse and base respectively?

(page 78.)

	ft.	in.		fath	ft.	in.
Answer, Base	4	3·8326	Hyp.	1	1	4·7165

* In this example, as the angle exceeds 45°, it will be found standing on the right hand side of the page (as already explained), and the denomination of the required sides will be found at the bottom. A little attention to this order will prevent the mistake, which may otherwise take place, by an inversion of the base and perpendicular.

EXAMPLE.

Given the angle 59° 30′, perpendicular 1 fathom, the other sides are required.

(page 81.)

	fath.	ft.	in.			fath.	ft.	in.
Answer, Base	1	4	2·2317		Hyp.	1	5	9·8612

TABLE III.—EXAMPLE.

Given the angle 5°, base 1 fathom, the hypothenuse and perpendicular are required.

(page 83.)

	fath.	ft.	in.			fath.	ft.	in.
Answer, Hyp.	11	2	10·10734		Perp.	11	2	6·96374

EXAMPLE.

Given the angle 30°, base 1 fathom, the other sides are required.

(page 83.)

	fath.	ft.	in.			fath.	ft.	in.
Answer, Hyp.	2	0	0		Perp.	1	4	4·70766

NOTE.—The foregoing examples serve only to exemplify the manner of taking out the primes from the tables; and as the given side is exactly one fathom, of course the tables give a direct answer. In the following examples the mode of taking out the tabular numbers is precisely as the foregoing, but the number of fathoms contained in the length of the given side, will be the multiplier of the other side of the triangle.

PLANE TRIGONOMETRY.

BY THE TABLES.

—◆—

CASE I.

WHEN THE HYPOTHENUSE IS GIVEN.

RULE.—Look in the first table, and against the given angle stands the base and perpendicular, answering to one fathom of the hypothenuse; take out these numbers and multiply them respectively by the length of the hypothenuse.

EXAMPLE.

Given the angle 23° 30′, and hypothenuse 12 fathoms; the base and perpendicular are required.

OPERATION.

BASE.	PERPENDICULAR.
Feet 2 . 4·70993	Feet 5 . 6·02833
12	12
28 . 8·51916	66 . 0·33996

BY CONSTRUCTION.

PROCESS.   SCALE—40 *feet to an inch.*

Draw the line A B of any length, make the angle C = 23° 30′ by a scale of chords, or with a protractor; draw the hypothenuse A C = 72 feet from a scale of equal parts. From C let fall the perpendicular C B; then A B C is the triangle required. A B, measured by the same scale of equal parts, will be 28 feet 8½ inches, and B C will be 66 feet.

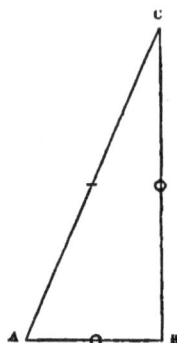

CASE II.

WHEN THE PERPENDICULAR IS GIVEN.

RULE.—Look in the second table, and opposite the given angle will be found the base and hypothenuse corresponding to one fathom of the perpendicular; multiply these numbers separately by the length of the perpendicular.

EXAMPLE.

Given the angle 39° 30′, and perpendicular 9 fathoms 3 feet; the hypothenuse and base are required.

OPERATION.

	fath.	ft.	in.
3 ¦ ½	0	4	11·3522
			9
	7	2	6·1698
	0	2	5·6761
Base	7	4	11·8459*

	fath.	ft.	in.
3 ¦ ½	1	1	9·3096
			9
	11	3	11·7864
	0	3	10·6548
Hyp.	12	1	10·4412

* It has been before observed that it would be better to bring the answer out in feet than in fathoms, as in the last case.

PROCESS.

Draw the line A B of a sufficient length, at any point
B erect the perpendicular
B C, which make equal to
57 feet by a scale of equal
parts.   At C make the
angle = 39° 30′, the com-
plement of A.   From C
draw the hypothenuse, and
it will cut the base A B in
the point A; then will A B
measure 47 feet, and A C
73 feet 10 inches.

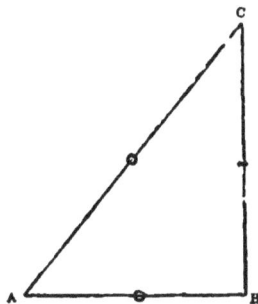

CASE III.

WHEN THE BASE IS GIVEN.

RULE.—Look in the third table, and opposite the
given angle (as in the former cases) the corresponding numbers to one fathom of base will be
seen, which, being multiplied by the given length
of the base, produces the hypothenuse and perpendicular.

EXAMPLE.

Given the angle 20 degrees, and base 28 feet 9
inches; the hypothenuse and perpendicular are
required.

		ft.	in.			ft.	in.
∠ 20°	½	17	6·51392*		½	16	5·81837*
			4				4
		70	2·05568			65	11·27348
	⅓	8	9·25696		⅓	8	2·90918
	⅛	2	11·08565		⅛	2	8·96972
		2	2·31424			2	0·72729
Hyp.		84	0·71253	Perp.		78	11·87967

BY CONSTRUCTION.

PROCESS.

Draw the base A B, which make = 28 feet 9 inches, from a scale of equal parts, at B erect the perpendicular B C, make the angle A = 70° and draw the hypothenuse A C to cut the perpendicular B C in the point C; then will A C measure 84 feet, and B C 78 feet 11½ inches.

* These numbers stand in the tables in fathoms, &c.; the hypothenuse will be found 2 fathoms 5 feet 6 inches, &c., and the perpendicular 2 fathoms 4 feet 5 inches, &c.

APPLICATION OF THE TABLES

TO

DIAGONAL SHAFTS.

—◆—

REMARKS.

As in the foregoing cases each side of the triangle is distinctly made radius, it follows that every problem in oblique dialling, &c., can be solved by one or the other of these cases; because, in every instance, a side and the angles are always given.

GENERAL RULE.

When the hypothenuse is given, work by case the first.

When the perpendicular is given, work by case the second.

When the base is given, work by case the third.

EXAMPLE 1.

A diagonal shaft A B was found to measure 84 feet,* and the angle of declination observed to

---

* When the given line is denominated in feet, it must be brought into fathoms by dividing it by 6 (the number of feet in a fathom); thus in the above example the shaft being 84 feet is 14 fathoms, and therefore the numbers are multiplied by 7 and 2, which are equal to 14.

be 48 degrees; required the base B C, and perpendicular A C.

BY CASE I.

```
         ft.  in.
  ∠ 48°  4   5·50643
              7
        ───────────
         31   2·54501
              2
        ───────────
  Base   62   5·09002


         ft.  in.
          4   0·17740
              7
        ───────────
         28   1·24180
              2
        ───────────
  Perp. 56   2·48360
```

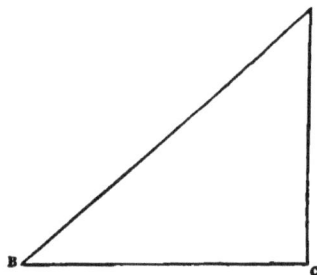

EXAMPLE 2.

A perpendicular shaft B C, measuring 57 feet,
was found to intersect an underlaying shaft A C,
whose angle of acclivity was observed to be 50° 30′;
required the length of the underlaying shaft A C,
and the distance from the perpendicular at the surface A B.

BY CASE II.

$\angle$ 50° 30'⎫
  Comp. ⎬ Base
39° 30'⎭

		ft.	in.
Base	½	4	11·3522
			9
		44	6·1698
		2	5·6761
A B		46	11·8459

		ft.	in.
Hyp.	½	7	9·3096
			9
		69	11·7864
		3	10·6548
A C		73	10·4412

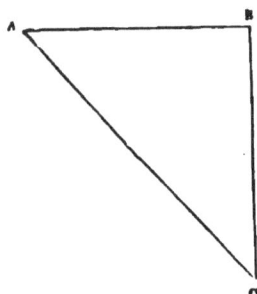

NOTE.—In the above example, the angle having again been taken with the horizon, the operative angle will be 39° 30', because 50° 30'—90°=39° 30'. We may also observe that, the length of the shaft being 57 feet, the multiplier is 9½, or 9 fathoms 3 feet.

EXAMPLE 3.

A horizontal cross-cut B C from the foot of a diagonal B A to a perpendicular shaft C A was found to measure 224 feet 8 inches, and the angle of acclivity (taken at B, the foot of the shaft) 40 degrees; I require the respective lengths of the hypothenuse A B and perpendicular A C.

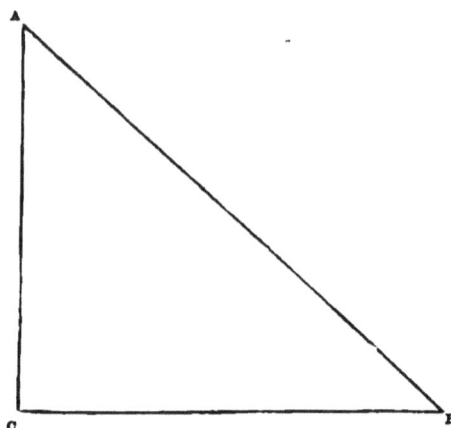

BY CASE III.

		ft.	in.			ft.	in.
∠ 40°	⅓	7	9·98932*		⅓	5	0·41517
Comp.			12				12
50°		93	11·87184			60	4·98204
			3				3
		281	11·61552			181	2·94612
		7	9·98932			5	0·41517
	⅓	2	7·32977		⅓	1	8·13889
		0	10·44325			0	6·71279
A B		293	3·37786	A C		188	6·21247

EXAMPLE 4.

When a lode has changed its underlay.

RULE.—Take out the numbers opposite the given angles, and work them by the former cases; then add their sums together respectively for the answer.

---

* It will be observed that this number stands in the table 1 fath. 1 ft. 9 98932 in., and the angle having been taken at the foot of the shaft, the complement of that angle (i.e., what it wants of 90°) must be used; therefore the above tabular numbers will be found in the column opposite 50°, being the complement of 40°.

PROBLEM.

In dialling a shaft sunk on a lode, it was found that the first draft B D measured 71 feet, on an

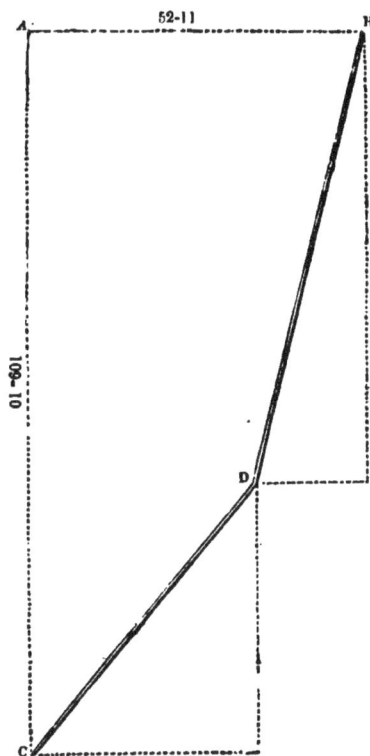

angle of 14° 45′, but from that depth to the foot of the shaft C the angle proved to be 40° 15′, and the length D C 54 feet; required the distance from the brace of the diagonal B, where a perpendicular shaft ought to be sunk, in order to come down exactly at the foot of the underlay; also the depth of the perpendicular A C.

OPERATION.

BASE.		PERPENDICULAR.

fath. ft. in.   fath. ft. in.

∠ 14° 45′ | 1 | 1/6 | 0   1   6·33134   1 | 1/6 | 0   5   9·6273
                                    12   12
              3   0   3·97608   11   3   7·5276
              0   0   3·05522   0   0 11·6045
              3   0   0·92086   11   2   7·9231

BASE.	PERPENDICULAR.

fath. ft. in.   fath. ft. in.

∠ 40° 15′ = 0   3   10·52093   0   4   6·95274
                              9   9
            5   4   10·68837   6   5   2·57466

SUMMARY OF BASES.   SUMMARY OF PERPENDICULARS.

fath. ft. in.   fath. ft. in.

  3   0   0·92086   11   2   7·9231
  5   4 10·68837   6   5   2·57466
  8   4 11·60923   18   1 10·49776
  6   6
A B 52 feet 11 in.   A C 109 feet 10 in.

EXAMPLE 5.

When a lode has changed or reversed its under-lay from north to south, or east to west.

RULE.—Add all the perpendiculars together, as in the last problem, but subtract the bases, made by the reverse or contrary shafts, one from the other; the remainder will be the true length of the base.

PROBLEM.

A diagonal shaft was found to incline and measure as follows, viz. :—

A B . . 54 feet 18° 45′
B C . . 42 do. 12° 15′
C D . . 69 do. 25° 0′

throughout the above drafts, the declination or

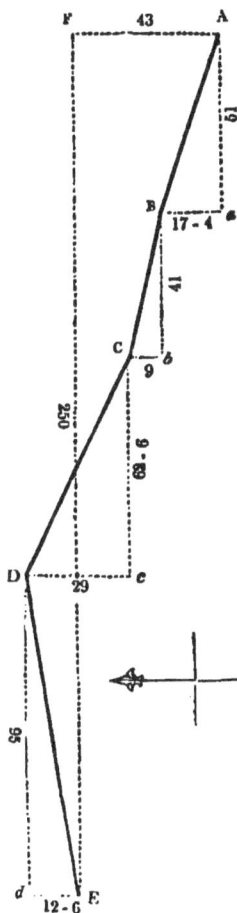

D

underlay bore northerly, but from that depth D it made an angle of 7° 30′ in a southerly direction, and this last draft D E measured 96 feet. It is required to know the perpendiculars and bases of all the foregoing sides respectively and collectively.

OPERATION.

BASES NORTHERLY.

PERPENDICULARS.

fath. ft. in.

∠ 18° 45′=0   1   11·14364
                              9

B a 2   5   4·29276

fath. ft. in.

0   5   8·17897
                  9

A a 8   3   1·61073

∠ 12° 15′=0   1   3·27680
                            7

C b 1   2   10·93760

0   5   10·36062
                    7

B b 6   5   0·52434

∠ 25° 0′=0   2   6·42852
                            11

4   3   10·71372
0   1   3·21426

D c 4   5   1·92798

0   5   5·25416
                  11

9   5   9·79576
0   2   8·62708

C c 10   2   6·42284

BASE SOUTHERLY.

fath. ft. in.

∠ 7° 30′=0   0   9·39789
                          8

1   0   3·18312
                2

E d 2   0   6·36624

fath. ft. in.

0   5   11·38403
                    8

7   5   7·07224
                2

D d 15   5   2·14448

· SUMMARY.

BASES.			PERPENDICULARS.		
fath.	ft.	in.	fath.	ft.	in.
2	5	4·29276	8	3	1·61073
1	2	10·93760	6	5	0·52434
4	5	1·92798	10	2	6·42284
North 9	1	·5·15834	15	5	2·14448
South 2	0	6·36624	41	3	10·70239
7	0	10·79210	6		
6			F E 249 ft. 10 in.		

F A 42 ft. 10 in.

EXAMPLE 6.

When a shaft has been sunk in error,* or not exactly at right angles with the lode.

RULE I.

Work for the base and perpendicular as before, by Case I.; then find the deviation by the following

RULE II.

Take out the base from the second table, standing opposite the angle of error, and multiply it by the length of the shaft.

* Underlaying shafts are always intended to be sunk at right angles with the lode ; that is, if the lode runs east and west, the horizontal bearing of the shaft will be either north or south, as the lode may happen to underlie. But it is sometimes the case, that through inattention of workmen or other causes, the shaft has declined from its true course and inclined toward the right or left ; and as this is neither a trivial nor uncommon occurrence, and admits not of development by the ordinary mode of dialling, we have here introduced a rule which will hold good in all cases of the kind.

PROBLEM.

An oblique shaft A B was found to measure 89 feet 6 inches, on an angle of 53° 15′, and it was also observed that the shaft had declined 3° 45′ west from the intended right angle of the east and west lode : required the base C D and perpendicular C A, and how far the shaft has departed from its true course A D.

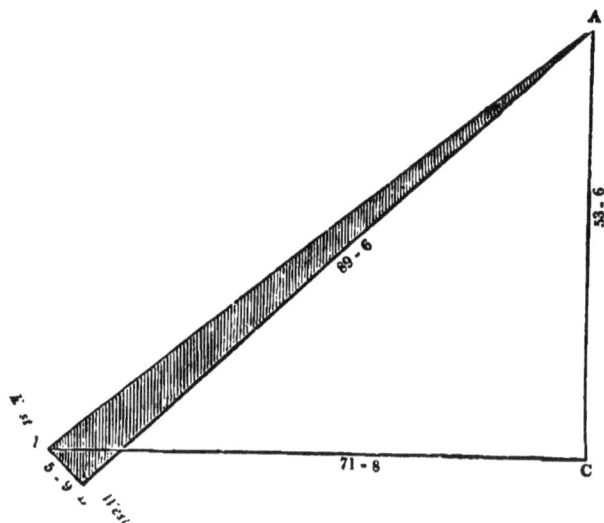

OPERATION.

			ft.	in.				ft.	in.
∠ 53° 15′	3	$\frac{1}{2}$	4	9·69027		3	$\frac{1}{2}$	3	7·07937
				7					7
			33	7·83189				25	1·55559
				2					2
			67	3·66378				50	3·11118
			2	4·84513				1	9·53968
	2	$\frac{1}{3}$	1	7·23009		2	$\frac{1}{3}$	1	2·35979
	0	6 $\frac{1}{4}$	0	4·80752		0	6 $\frac{1}{4}$	0	3·58994
	Base	71		8·54652		Perp.	53		6·60059

THEN FOR THE DEVIATION.

		ft.	in.
Table 2nd.—∠ 3° 45′ Base	½	0	4·7189
			7
		2	9·0323
			2
		5	6·0646
		0	2·3594
	⅓	0	0·78
	¼	0	0·19
		5	9·3940

Thus it is clear that if the above shaft were sunk on an east and west lode and the angle of error inclined westerly, that the foot of the shaft B would be 5 feet 9⅓ inches in that direction beyond its designed course A D.

EXAMPLE 7.

To find the perpendicular depth of the junction of lodes.

CASE I.

When two lodes underlay in the same direction.

RULE.—Subtract the tabular number of the base of the lesser angle from the greater; then by direct proportion, say,

As this difference
Is to one fathom perpendicular,
So is the distance of the lodes at surface
To the junction of the lodes.

PROBLEM.

Two lodes were discovered at the surface, 12 fathoms apart from C to D, both underlaying north. The southernmost lode D B made an angle of 38° 15′: the other C B 23°. Required the perpendicular depth A B where these lodes will unite, supposing they both regularly continue their respective angles of declination.

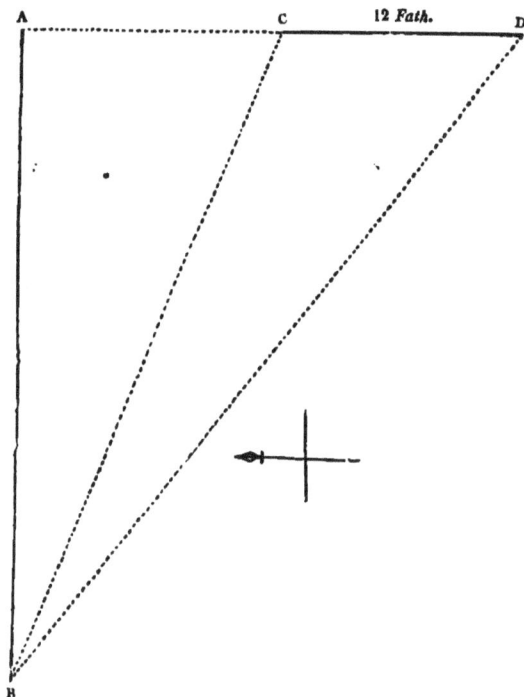

OPERATION

$$\begin{array}{lll} & \text{ft.} & \text{in.} \\ \text{From} \angle 38° \ 15' = & 4 & 8 \cdot 7602 \\ \text{Take} \angle 23° \ \ 0' = & 2 & 65 \cdot 622 \\ \hline & 2 & 2 \cdot 1980 \end{array}$$

```
            ft.  in.      fath.     fath.
Then, As 2   2·198   :  1   ::   12
        12                        6
        ──────                    ──
        26·198                    72
                                  12
                                  ──
                     26·198) 864·000 (32·9
                             78594        6
                             ─────       ──
                             78060      5·4
                             52396       12
                             ──────     ───
                             256640     4·8
                             235782
                             ──────
                              20858
```

```
        fath.  ft.  in.
Answer   32    5    4 A B
```

DIAGONAL LODES.

—◦—

IF it is required to find the respective lengths of the lodes C B and D B, and the horizontal line D A, work by Case II., where the perpendicular is given.

TO FIND D B.

fath.	ft.	in.
∠ 38° 15′ ⅑ │ 1	1	7·6831
		11
14	0	0·5141
		3
42	0	1·5423
0	0	10·1870
D B 41	5	3·3553

Here we multiply by 33 and subtract from the product what the hypothenuse is minus of that measure, which, being eight inches, is one-ninth of a fathom. This is the shortest method.

TO FIND C B AND D A, OR C A.

BASE.

fath.	ft.	in.
∠ 23° 0′ ⅑ │ 0	2	6·5622
		11
4	4	0·1842
		3
14	0	0·5526
0	0	3·3964
C A 13	5	9·1580

HYPOTHENUSE.

fath.	ft.	in.
⅑ │ 1	0	6·2179
		11
11	5	8·3969
		3
35	5	1·1907
0	0	8·6908
C B 35	4	4·4999

Then C A + C D = D A, or C A added to C D gives the line DA, 25 fathoms 5 feet 9 inches, &c.

EXAMPLE 8.

To find the perpendicular depth of the junction of lodes.

CASE II.

When two lodes, by their underlay, incline indirectly towards each other.

RULE. — Add the tabular bases together, then find the depth by direct proportion as in the last example.

PROBLEM.

Two lodes were observed 36 fathoms apart at the surface from A to C, the northernmost lode A underlaying south 18° 15′, and the southernmost lode C underlaying north 31° 45′; required the

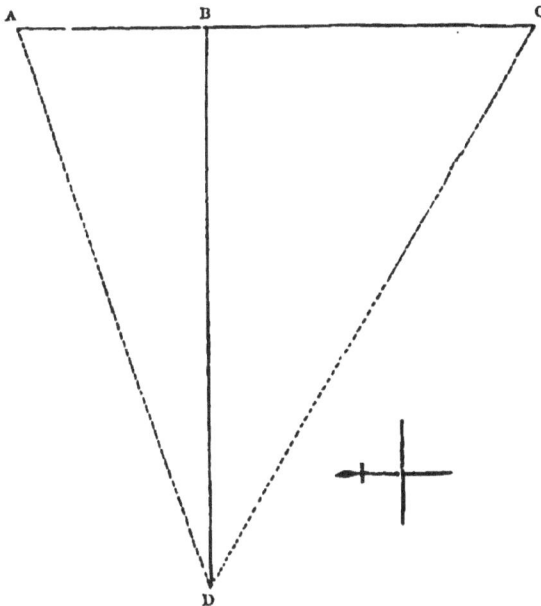

depth B D at which these lodes will intersect each other.

$$
\begin{array}{lll}
 & \text{ft.} & \text{in.} \\
\text{To } \angle 18^\circ\ 15' = 1 & & 11\cdot6220 \\
\text{Add} \angle 31^\circ\ 45' = 3 & & 8\cdot5550 \\
\hline
5 & & 8\cdot177
\end{array}
$$

$$
\begin{array}{ccc}
\text{ft.} & \text{in.} & \text{fath. fath.} \\
\text{Then, As } 5 & 8\cdot177 & : 1 :: 36 \\
12 & & 6 \\
\hline
68\cdot177 & & 216 \\
& & 12 \\
\hline
\end{array}
$$

$$
68\cdot177)\ \overline{2592\cdot0000}\ (38\cdot0 \\
204531 \\
\hline
546690 \\
545416 \\
\hline
12740
$$

Answer 38 fathoms.

If required to find the length of the lodes A D and C D and the distance of the shaft B from the lodes C and A at the surface, work by Case II. thus :

BASE.			HYPOTHENUSE.		
fath.	ft.	in.	fath.	ft.	in.
$\angle 18^\circ\ 15' = 0$ 1	11·622		1	0	3·8135
		6			6
1	5	9·732	6	1	10·8810
		6			6
11	4	10·392	37	5	5·2860
0	3	11·244	2	0	7·6270
A B 12	2	9·636	A D 40	0	0·9130

TO FIND B C AND C D.

BASE.				HYPOTHENUSE.		
	fath.	ft.	in.	fath.	ft.	in.
∠ 31° 45′ = 0	3	8·555		1	1	0·7008
		6				6
	3	4	3·330	7	0	4·2048
			6			6
	22	1	7·980	42	2	1·2288
	1	1	8·110	2	2	1·4016
B C *	23	3	1·090	C D 44	4	2·6304

* It may be observed that A B and B C added together do not make 36 fathoms, by something more than an inch: now this does not happen through any defect in the tables, but because the perpendicular has not been worked out—for if the remainder (12740) were prosecuted, the perpendicular would prove to be 38 fath. 0 ft. 1·3392 in., instead of 38 fathoms, which addition to multipliers would make up the exact deficiency.

PERPENDICULAR SHAFTS

AND

LEVELS.

— ✦ —

RULE.—When the angle of acclivity is given, take the complement (or what it wants of 90°) for the operative angle; in every other particular, work by the former cases.

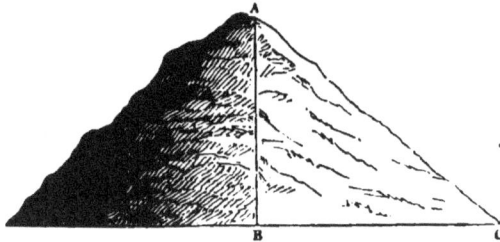

EXAMPLE 1.

A perpendicular shaft having been sunk from the top of a hill at A, from whence the slope to C measured 330 feet: It is required to know the length an adit must be driven from the base of the hill at C to intersect the shaft at B, and what will be the depth of the shaft at that intersection, the angle of acclivity at C being 41 degrees.

BY CASE II.

	PERPENDICULAR.			BASE.	
	ft.	in.		ft.	in.
Comp. of 41° is 49°	3	11·23625		4	6·33909
		9			9
	35	5·12625		40	9·05181
		6			6
	212	6·75750		244	6·31086
	3	11·23625		4	6·33909
	216	5·99375		249	0·64995

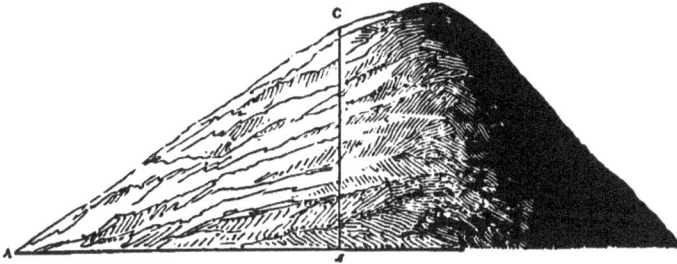

EXAMPLE 2.

An adit having been driven 75 fathoms from A to B, required to know how far up the hill from A I ought to measure in order that a perpendicular may be sunk to intersect the adit at x, 58 fathoms from the tail at A; also the depth of the shaft C x, the angle of acclivity from A towards C being 33 degrees. Or thus : — Given the base 58 fathoms ; angle of acclivity 33°, of which the complement or angle of declivity is 57 °; required the hypothenuse and perpendicular.

BY CASE III.

	HYPOTHENUSE.			PERPENDICULAR.	
	ft.	in.		ft.	in.
Comp. ∠ 33° is 57°=	7	1·85016		3	10·75735
		8			8
	57	2·80128		31	2·05880
		7			7
	400	7·60896		218	2·41160
	14	3·70032		7	9·51470
A C	414	11·30928	C x	225	11·92630

EXAMPLE 3.

From the foot of a perpendicular shaft A B, 70

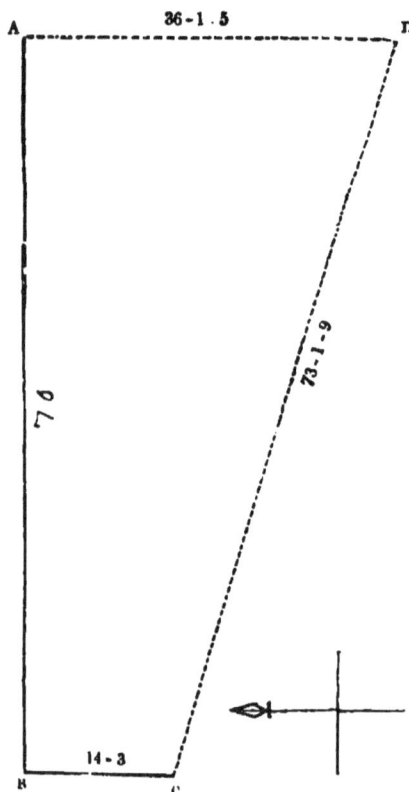

fathoms in depth, a cross-cut was driven south 14 fathoms 3 feet in length (C), where a lode was discovered underlaying north, and the angle of ascension or elevation 72° 45′: required the length of this lode from the end of the drift C to the surface D: also the distance from the brace of the perpendicular shaft A to the back of the lode at grass (D), supposing the lode to have a regular underlay.

BY CASE II.

	BASE.			HYPOTHENUSE.		
	fath.	ft.	in.	fath.	ft.	in.
Comp. of ⎱ ∠ 72° 45′ is 17° 15′ ⎰	0	1	10·3560	1	0	3·3911
			7			7
	2	1	0·4920	7	1	11·7377
			10			10
	21	4	4·9200	73	1	9·3770
Add length of drift	14	3	0			
	36	1	4·92			

ANSWER.

	fath.	ft.	in.
Length of lode	73	1	9¼
Distance from shaft ⎱ at the surface ⎰	36	1	5

EXAMPLE 4.

From the depth of 36 fathoms 4 feet, in an engine shaft A B, a cross-cut was driven which pierced a lode C, 14 fathoms 2 feet from B. The lode was found to make an angle of 30 dgrees, inclining towards the shaft. Required the depth at which the shaft will intersect the lode, and the length of the lode from C to the point of intersection o.

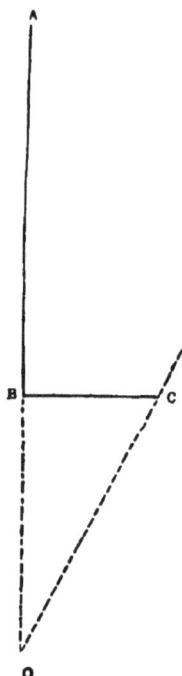

BY CASE III.

	HYPOTHENUSE.		
	fath.	ft.	in.
∠ 30° ⅓)	2	0	0
			2
	4	0	0
			7
	28	0	0
	0	4	0
	28	4	0

	PERPENDICULAR.		
	fath.	ft.	in.
⅓)	1	4	4·70766
			2
	3	2	9·41532
			7
	24	1	5·9072
	0	3	5·569
	24	4	11·47624

ANSWER.

	fath.	ft.	in.
Depth from A to B	36	4	0
Depth from B to o	24	4	11·47624
Extreme depth	61	2	11·47624

	fath.	ft.
Length from C to o	28	4

SLIDES.

—+—

WHEN a lode has been thrown up by a slide, to find the base and perpendicular.

RULE.—Add the bases made by the segments of the lode together for the horizontal, and subtract the perpendicular made by the accession of the slide from the sum of the others for the perpendicular.

EXAMPLE.

A shaft having been sunk on a lode 114 feet from A to B, on an angle of 54° 30′, at this place the lode was separated and thrown up by a slide, from B to C, 32 feet, the angle of elevation at B being 47°: at C the lode was again cut and prosecuted on an angle of 51°, from C to D, 73 feet. Required to know the length from A to E at surface, where a perpendicular shaft should be put down, that would intercept the lode at the foot of the diagonal C D; also the depth of the shaft E D.

E

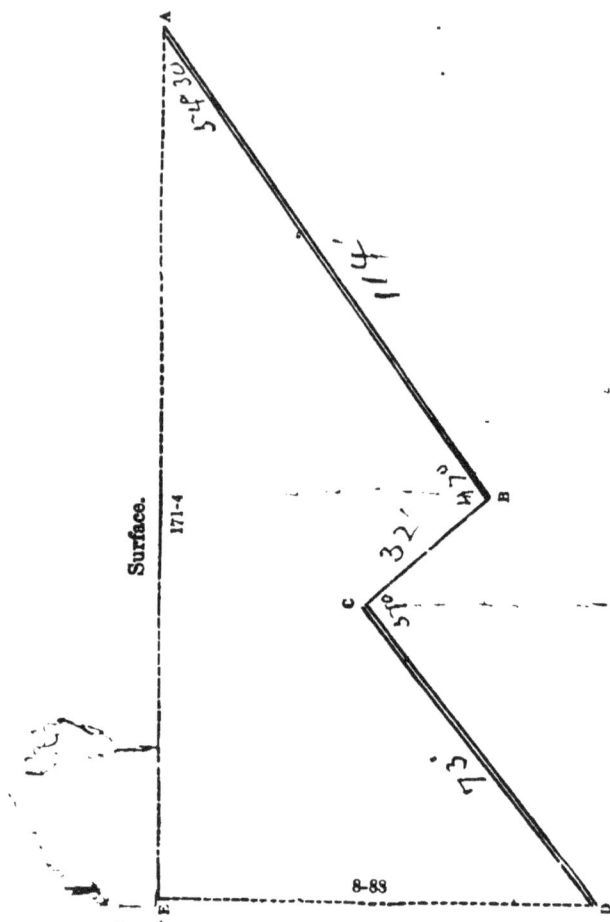

THE FOREGOING EXAMPLE BY THE TABLES.

	BASE.		PERPENDICULAR.		
	ft.	in.	ft.	in.	
∠ A 54° 30′ =	4	10·61632	3	5·81062	
		6		6	
	29	3·69792	20	10·86372	(Multiplier
		3		3	19 faths.)
	87	11·09376	62	8·59116	
	4	10·61632	3	5·81062	
	92	9·71008	66	2·40178	

		BASE.			PERPENDICULAR.

		ft. in.			ft. in.
∠ B 47° 0′ ⎫ Comp. 43° 0′ ⎬ =	⎮⅓⎮	4 1·10388 5	⎮⅓⎮	4 4·65747 5	
		20 5·51940 1 4·36796		21 11·28735 1 5·552	(Multiplier 5 fath. 2 ft.)
		21 9·88736		23 4·83935	

		ft. in.			ft. in.
∠ C .51° 0′	⎮⅙⎮	4 7·95451 12	⎮⅙⎮	3 9·31107 12	
		55 11·45412 0 9·32575		45 3·73284 0 7·55184	(Multiplier 12 fa. 1 ft.)
		56 8·77987		45 11·28468	

SUMMARY.

BASES.	PERPENDICULARS.
ft. in.	ft. in.
92 9·71008	∠ A= 66 2·40178
21 9·88736	∠ C= 45 11·28468
56 8·77987	112 1·68646
Answer E A = 171 4·37731	∠ B = 23 4·83935
	D E = 88 8·84711

NOTE.—Perpendiculars to strike B or C and their respective horizontal distances from A are shown by the sums of the first and second operation in the above calculations.

When a lode has been thrown down by a slide.

RULE.—Add the perpendiculars together for the depth, and subtract the base of the slide from the bases of the segments of the lode for the horizontal.

EXAMPLE.

A shaft A B having been sunk 77 feet on a lode, which made an angle of 34° 45′, it was there found that a slide had severed or disjointed the lode and

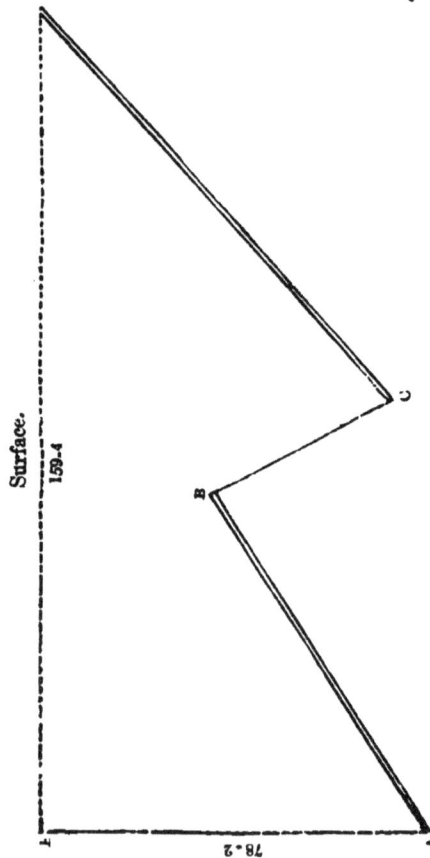

carried it downward from B to C 40 feet, on an angle of depression or declivity 59°. Here (at C) the lode was again discovered, and wrought 102 feet from C to D, on an angle of 42° 15′; required the depth of the vertical line D E, and length of the horizontal A E.

NOTE.—Should it be required to find the proper depth in the shaft D E from whence to drive a cross-cut to strike the end of the shaft A B (where the slide first appeared)—the perpendicular of the first draft gives the depth, and the length of the cross-cut will be found by subtracting the base of the first draft from the horizontal line A E; and should it be necessary to make a drift from the shaft D E to the angle C (where the lode was again discovered), the depth will be found by adding the perpendiculars of A and B together, and the bases of B and C will be the length of the drift.

THE FOREGOING EXAMPLE BY THE TABLES.

			BASE.			PERPENDICULAR.
			ft. in.			ft. in.
∠ A 34° 45′	$\frac{1}{2}$	3	5·03977	$\frac{1}{2}$	4	11·15858
			12			12
		41	0·47724		59	1·90296
		1	8·51988		2	5·57929
	$\frac{1}{3}$	1	1·67992	$\frac{1}{3}$	1	7·71952
		43	10·67704		63	3·20177

			ft. in.			ft. in.
∠ B 59° 0′	$\frac{1}{2}$	5	1·71605	$\frac{1}{2}$	3	1·08274
			·6			6
		30	10·29630		18	6·49644
	$\frac{1}{3}$	2	6·85802	$\frac{1}{3}$	1	6·54137
		0	10·28600		0	6·18045
		34	3·44032		20	7·21826

BASE.			PERPENDICULAR.	
	ft.	in.	ft.	in.
∠ C 42° 15′	4	0·41041	4	5·2957
		4		4
	16	1·64164	17	9·1828
		4		4
	64	6·56656	71	0·7312
	4	0·41041	4	5·2957
	68	6·97697	75	6·0269

SUMMARY.

	BASES.		PERPENDICULARS.	
	ft.	in.	ft.	in.
∠ A . . .	43	10·67704	63	3·20177
∠ C . . .	68	6·97697	20	7·21826
	112	5·65401		
∠ B . . .	34	3·44032	75	6·0269
Ans. A E =	78	2·21369	D E=159	4·44693

.

HORIZONTAL DIALLING.

RULE.—Observe which side of the triangle is given, and work by the specified case.

When there is more than one draft in the operation, add the sums of the respective sides together for the answer.

EXAMPLE 1.

Being required to put down a shaft 618 feet due east of an engine shaft at A, I am prevented from

measuring in a direct line by intervening hills and wood: I therefore find it necessary, in order to

avoid these obstructions, to go on an angle of 27°
south of east from the shaft A. What distance
must I proceed in this direction before I come at
right angles with, or due south of, the eastern extremity of the given line, and how far must I then
measure in a northerly direction to come exactly
on the required spot? Or the question may stand
thus:—Given the perpendicular 618 feet, angle
27°; the hypothenuse and base are required.

OPERATION.

		BASE.		HYPOTHENUSE.	
		ft.	in.	ft.	in.
27°=		3	0·6858	6	8·8075
			10		10
		30	6·8580	67	4·0750
			10		10
(Multiplier 103 fath.)		305	8·5800	673	4·7500
		9	2·0574	20	2·4225
		314	10·6374	693	7·1425

EXAMPLE 2.

It being required to find the distance between
two shafts A and B, which are inaccessible to a
direct measurement on account of a marsh or lake
lying in the way: I consequently measure 352 feet
on an angle of 63° south of west, from A to C; at
this station (C) I can see the shaft B, which I find
by observation bears 29° north of west, and the line
from C to B measures 615 feet; how far are these

shafts apart in a right line? Or the question may stand simply thus: —

$$\left\{ \begin{array}{l} \text{Given angle } 63°, \text{ hypothenuse } 352 \text{ feet.} \\ \text{Given angle } 29°, \text{ hypothenuse } 615 \text{ feet.} \end{array} \right\}$$

The sum of the perpendiculars* is required.

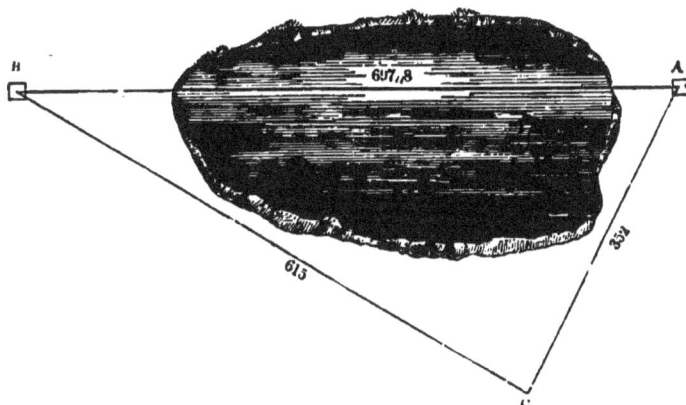

OPERATION.

PERPENDICULAR.		PERPENDICULAR.	
ft.	in.	ft.	in.

∠ A 63° | ½ | 2   8·68732   ∠ C 29° | ½ | 5   2·97262

(Multiplier 58 fa. 4 ft.)

```
 ∠ A 63° │½│   2   8·68732     ∠ C 29° │½│   5   2·97262
         │ │        11                 │ │       10
         │ │  29  11·56052             │ │  52   5·72620
         │ │        5                  │ │       10
         │ │ 149   9·80260             │ │ 524   9·26200
         │ │   8   2·06196             │ │  10   5·94524
         │ │   1   4·34366             │ │   2   7·48631
         │⅓│   0   5·44788             │ │ 537  10·69355
         │ │ 159   9·65610
```

(Multiplier 102 fa. 3 ft.)

SUM.

	ft.	in.
	159	9·65610
	537	10·69355
Ans. A B.	697	8·34965

* As in this instance the perpendicular only is wanted, there is no necessity for taking out the other side.

VERTICAL DIALLING;

OR THE

MENSURATION OF HEIGHTS.

—————

RULE.— Observe the given side and angle, and work by the respective cases as heretofore.

EXAMPLE 1.

From the bottom of a tower at B, I measured 200 feet in a direct line B A on an horizontal plane; I then took the angle A 42° : required the height of the tower and staff B C.

OPERATION.

ft. in.

Complement of ∠ A ⎱
42° is ∠ C 48° ⎰ | ⅓ | 5   4·82909
                                     11

                                 59   5·11999
                                          3

(Multiplier)   178   3·35997
(53 fath. 2 ft.)   1   9·60969

Ans. B C 180   0·96966

In operations of this nature the hypothenuse need not be regarded.

EXAMPLE 2.

Wanting to ascertain the height of an irregular hill, I proceed, from the several stations A, B, and C, to take the angles and measure the distances as follows, viz. :

From A to B ∠ 41°   0′ length 210 feet.
From B to C ∠ 22°   0′ length 216 feet.
From C to D ∠ 37° 30′ length 247 feet.

Required the altitude E D.

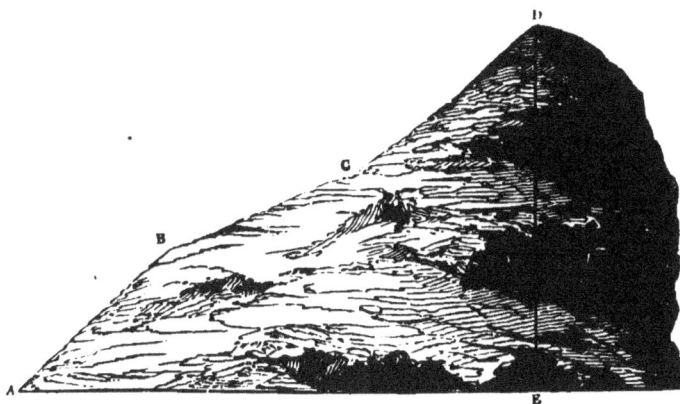

OPERATION.

PERPENDICULAR.

ft.   in.

∠ A 41° 0′ Comp. 49° 0′   3 11·23625
                                        7
6)210 feet   27   6·65375
 35 Multiplier   5
                               137   9·26875

       ∠ B 22° 0′ Comp. 68° 0′   2   2·97167
6)216   6
 36 Multiplier   13   5·83002
                                       6
                                80 10·98012

       ∠ C 37° 30′ Comp. 52° 30′ |⅙| 3   7·83082
                                       10
6)247   36   6·30820
 41 . 1 Multiplier   4
                                146   1·23280
                                  3   7·83082
                                  0   7·30513
                                150   4·36875

                    ft.   in.
               A 137   9·26875
               B 80 10·98012
               C 150   4·36875
Ans. height E D 369   0·61762

MISCELLANEOUS EXAMPLES IN THE FOREGOING RULES.

Given the hypothenuse 14 feet 5 inches, angle 88· : required the base and perpendicular.*

Answer $\left\{\begin{array}{l}\end{array}\right.$

	ft.	in.
Base =	14	4·89
Perp.=	0	6·03

Given the perpendicular 100 feet, angle 60° : required the hypothenuse and base.

Answer $\left\{\begin{array}{l}\end{array}\right.$

	ft.	in.
Hyp.=	200	
Base =	173	2·428

Given the base 118 feet, angle (Comp.) 23° : required the hypothenuse and perpendicular.

Answer $\left\{\begin{array}{l}\end{array}\right.$

	ft.	in.
Hyp.=	301	11·97
Perp.=	278	0·02

Given the angle 53° : required the underlay in a fathom.†

fath.	ft.	in.

Answer 1   1   11·5472

Given the angle 36° 45′ : required the underlay in a fathom.

ft.   in.

Answer 4   5·5650

---

* In single drafts, one or two figures of the decimal will be sufficient, the others may be rejected.

† It has been before observed, that the underlay is given in the base of the second table.

Given the angle 4° 15′ : required the underlay in a fathom.

in.

Answer 5·3496

A diagonal shaft having been sunk 8° 30′ out of its true course ; what will be the extent of departure, supposing the length of the shaft 76 feet?

ft.   in.

Answer 11   4·9577

Suppose a diagonal shaft were sunk as follows, viz. :

$$\angle\, 87°\ \ 0′ = 14\ \ 5$$
$$\angle\, 47\ \ \ \ 0\ = 11\ \ 2$$
$$\angle\, 87\ \ 30\ = 36\ \ 3$$
$$\angle\, 69\ \ 30\ = 26\ \ 2$$
$$\angle\, 77\ \ 30\ = 23\ \ 2$$
$$\angle\, 65\ \ 30\ = \ \ 9\ \ 2$$

Required the sum of the bases and perpendiculars.

Answer { Perp.   52   8·45944
         { Bases 169   6·90784

Wanting to know the distance between two shafts, inaccessible in a right line, I measured from the first shaft 126 feet, on an angle of 27° 15′ E. by N. ; from this station to the second shaft the line measured 91 feet, on an angle of 42° 30′ N. by W. : how far are the shafts apart ?

ft.   in.

Answer 179   1·3

Wanting to know the altitude of a precipice, I measured off from its base 66 feet, and from thence I take the angle to the summit, which I find to be 42° (and consequently the complement 48°): required the height.

ft.   in.

Answer 59   5·91999

At the foot of a hill the angle to the summit was 36°, from this place an adit had been driven in a direct line 218 feet: how far must I measure up the hill to put down a perpendicular shaft on the end of the adit, and what will be the depth of the shaft?

Answer $\begin{cases} \text{Hyp. or slope } 281 \quad 5\text{·}54804 \\ \text{Perp. or shaft } 158 \quad 4\text{·}62816 \end{cases}$

At the foot of a diagonal shaft, 28 fathoms in length, sunk on a lode 27° 45' underlaying north, another lode was cut making an angle 48° 45' underlaying south: what is the distance from the brace of the shaft to the back of the north lode?

FIRST TABLE.

	BASE.		PERPENDICULAR.	
	ft.	in.	ft.	in.
∠ 27° 45'	2	9·5	5	3·7
		7		7
	19	6·5	37	1·9
		4		4
	78	2·0	148	7·6

Divide this by 6 for the multiplier of the 2nd angle,

fath. ft. in.

which will be   24   4   7·6

THEN — SECOND TABLE.

BASE.

		ft.	in.
∠ 48° 45′	½	5	3·14
Comp.			6
41° 15′		31	6·84
			4
		126	3·36
		2	7·57
		0	10·52
		0	5·26
		0	0·65
		130	3·36

ANSWER.

	ft.	in.
Base of north lode	78	2
Base of south lode	130	3
Required distance	208	5

A perpendicular shaft having been sunk 168 feet in the side of a mountain, the slope or declivity making an angle with the shaft of 54° 15′: required to know how far I must measure down the hill to get at the right spot for driving an adit to come in the exact depth of the shaft,—the length of the adit is also required.

		ft.	in.
Answer	{ Slope	287	8·58
	{ Adit	233	4·40

A lode underlaying south was observed to make an angle of 17° 15′: required to know what distance from the back of the lode will be proper for sinking a perpendicular shaft that shall intersect the lode at the depth of 45 fathoms.

	ft.	in.
Answer	83	10·020

F

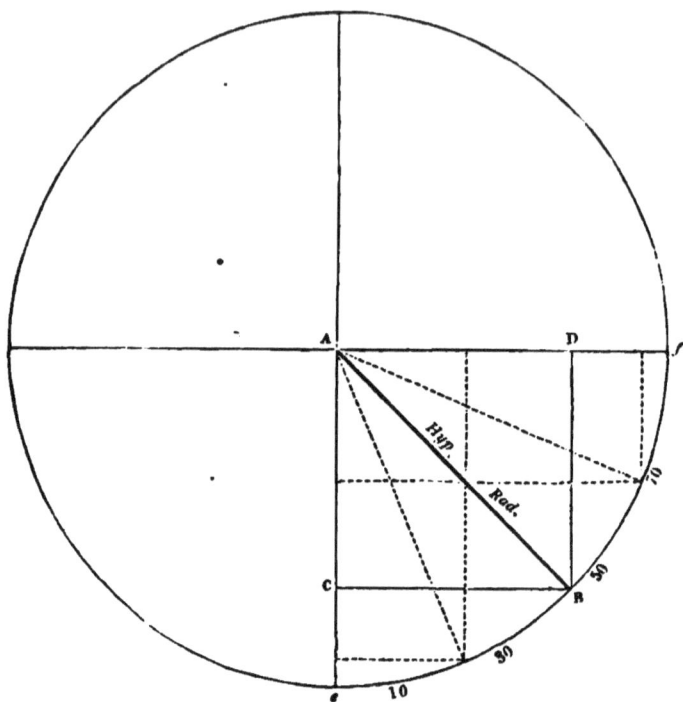

FIRST TABLE.

HYPOTHENUSE RADIUS,

ONE FATHOM.

ANGLE		BASE			PERPENDICULAR				
Deg.	Min.	Feet	Ins.	Decimals	Feet	Ins.	Decimals	Deg.	Min.
	1	0	0	·02094	6	0		89	59
	2	0	0	·04189	6	0		89	58
	3	0	0	·06283	6	0		89	57
	4	0	0	·08387	6	0		89	56
	5	0	0	·10482	6	0		89	55
	6	0	0	·12576	6	0		89	54
	7	0	0	·14670	6	0		89	53
	8	0	0	·16765	6	0		89	52
	9	0	0	·18859	6	0		89	51
	10	0	0	·20943·	6	0		89	50
	11	0	0	·23038	6	0		89	49
	12	0	0	·25132	6	0		89	48
	13	0	0	·27225	6	0		89	47
	14	0	0	·29319	6	0		89	46
	15	0	0	·31414	5	11	·99932	89	45
	30	0	0	·62831	5	11	·99726	89	30
	45	0	0	·94245	5	11	·99381	89	15
	PERPENDICULAR				BASE			ANGLE	

NOTE.—This page had no place in the former edition, but will be found useful in particular cases for long lines where the angle is required to be very minute. It will be seen that as there is but the thousandth part of an inch difference in one fathom between the hypothenuse and perpendicular on the first 15′ or first $\frac{1}{4}$ of a degree, the introduction of the decimal at any less fraction would be useless.

| ANGLE | | BASE | | | PERPENDICULAR | | | | |
Deg.	Min.	Feet	Ins.	Decimals	Feet	Ins.	Decimals	Deg.	Min.
1		0	1	·25657	5	11	·98903	89	
	15	0	1	·57067	5	11	·98286		45
	30	0	1	·88474	5	11	·97532		30
	45	0	2	·19877	5	11	·96664		15
2		0	2	·51276	5	11	·95614	88	
	15	0	2	·82666	5	11	·94249		45
	30	0	3	·14060	5	11	·93147		30
	45	0	3	·45442	5	11	·91708		15
3		0	3	·76819	5	11	·90132	87	
	15	0	4	·08188	5	11	·88420		45
	30	0	4	·39549	5	11	·86571	··	30
	45	0	4	·70902	5	11	·84584		15
4		0	5	·02246	5	11	·82461	86	
	15	0	5	·33581	5	11	·80201		45
	30	0	5	·64905	5	11	·77805		30
	45	0	5	·96219	5	11	·75272		15
5		0	6	·27521	5	11	·72602	85	
	15	0	6	·58811	5	11	·69793		45
	30	0	6	·90090	5	11	·66853		30
	45	0	7	·21354	5	11	·63773		15
6		0	7	·52605	.5	11	·60558	84	
	15	0	7	·83842	5	11	·57205		45
	30	0	8	·15063	5	11	·53718		30
	45	0	8	·46269	5	11	·50093		15
7		0	8	·77459	5	11	·46333	83	
	15	0	9	·08633	5	11	·42435		45
	30	0	9	·39789	5	11	·38403		30
	45	0	9	·70926	5	11	·34234		15
8		0	10	·02046	5	11	·29930	82	
	15·	0	10	·33147	5	11	·25490		45
	30	0	10	·64228	5	11	·20914		30
	45	0	10	·95288	5	11	·16203		15
9		0	11	·26328	5	11	·11356	81	
	15	0	11	·57347	5	11	·06374		45
	30	0	11	·88343	5	11	·01256		30
	45	1	0	·19316	5	10	·96004		15
10		1	0	·50267	5	10	·90616	80	

| PERPENDICULAR | BASE | ANGLE |

ANGLE		BASE			PERPENDICULAR			ANGLE	
Deg.	Min.	Feet	Ins.	Decimals	Feet	Ins.	Decimals	Deg.	Min.
10	15	1	0	·81193	5	10	·85093		45
	30	1	1	·12096	5	10	·79435		30
	45	1	1	·42973	5	10	·73643		15
11		1	1	·73825	5	10	·67716	79	
	15	1	2	·04650	5	10	·61654		45
	30	1	2	·35449	5	10	·55458		30
	45	1	2	·66221	5	10	·49128		15
12		1	2	·96964	5	10	·42663	78	
	15	1	3	·27680	5	10	·36062		45
	30	1	3	·58365	5	10	·29331		30
	45	1	3	·89021	5	10	·22464		15
13		1	4	·19648	5	10	·15465	77	
	15	1	4	·50243	5	10	·08331		45
	30	1	4	·80807	5	10	·01062		30
	45	1	5	·11338	5	9	·93663		15
14		1	5	·41838	5	9	·86129	76	
	15	1	5	·72204	5	9	·72304		45
	30	1	6	·02736	5	9	·70663		30
	45	1	6	·33134	5	9	·62730		15
15		1	6	·63497	5	9	·54666	75	
	15	1	6	·93825	5	9	·46469		45
	30	1	7	·24116	5	9	·38139		30
	45	1	7	·54371	5	9	·29677		15
16		1	7	·84589	5	9	·21084	74	
	15	1	8	·14769	5	9	·12359		45
	30	1	8	·44910	5	9	·03502		30
	45	1	8	·75014	5	8	·94514		15
17		1	9	·05076	5	8	·85395	73	
	15	1	9	·35099	5	8	·76143		45
	30	1	9	·65082	5	8	·66762		30
	45	1	9	·95023	5	8	·57250		15
18		1	10	·24922	5	8	·47607	72	
	15	1	10	·54779	5	8	·37833		45
	30	1	10	·84594	5	8	·27931		30
	45	1	11	·14364	5	8	·17897		15
19		1	11	·44091	5	8	·07734	71	
	15	1	11	·73772	5	7	·97441		45

PERPENDICULAR	BASE	ANGLE

ANGLE		BASE			PERPENDICULAR				
Deg.	Min.	Feet	Ins.	Decimals	Feet	Ins.	Decimals	Deg.	Min.
19	30	2	0	·03409	5	7	·87019		30
	45	2	0	·33000	5	7	·76467		15
20		2	0	·62545	5	7	·65787	70	
	15	2	0	·92043	5	7	·54977		45
	30	2	1	·21493	5	7	·44040		30
	45	2	1	·50895	5	7	·32973		15
21		2	1	·80249	5	7	·21779	69	
	15	2	2	·09554	5	7	·10457		45
	30	2	2	·38809	5	6	·99007		30
	45	2	2	·68013	5	6	·87429		15
22		2	2	·97167	5	6	·75724	68	
	15	2	3	·26270	5	6	·63892		45
	30	2	3	·55320	5	6	·51932		30
	45	2	3	·84320	5	6	·39847		15
23		2	4	·13264	5	6	·27635	67	
	15	2	4	·42156	5	6	·15297		45
	30	2	4	·70993	6	6	·02833		30
	45	2	4	·99776	5	5	·90243		15
24		2	5	·28503	5	5	·77528	66	
	15	2	5	·57176	5	5	·64686		45
	30	2	5	·85791	5	5	·51721		30
	45	2	6	·14350	5	5	·38631		15
25		2	6	·42852	5	5	·25416	65	
	15	2	6	·71295	5	5	·12077		45
	30	2	6	·99680	5	4	·98614		30
	45	2	7	·28006	5	4	·85027		15
26		2	7	·56272	5	4	·71317	64	
	15	2	·7	·84479	5	4	·57483		45
	30	2	8	·12622	5	4	·43528		30
	45	2	8	·40708	5	4	·29448		15
27		2	8	·68732	5	4	·15247	63	
	15	2	8	·96692	5	4	·00923		45
	30	2	9	·24590	5	3	·86478		30
	45	2	9	·52424	5	3	·71911		15
28		2	9	·80195	5	3	·57223	62	
	15	2	10	·07902	5	3	·42413		45
	30	2	10	.35543	5	3	·27483		30

PERPENDICULAR	BASE	ANGLE

ANGLE		BASE			PERPENDICULAR				
Deg.	Min.	Feet	Ins.	Decimals	Feet	Ins.	Decimals	Deg.	Min.
28	45	2	10	·63120	5	3	·12433		15
29		2	10	·90630	5	2	·97262	61	
	15	2	11	·18073	5	2	·81972		45
	30	2	11	·45450	5	2	·66561		30
	45	2	11	·72759	5	2	·51031		15
30		3	0	·00000	5	2	·35383	60	
	15	3	0	·27173	5	2	·19616		45
	30	3	0	·54276	5	2	·03730		30
	45	3	0	·81310	5	1	·87726		15
31		3	1	·08274	5	1	·71605	59	
	15	3	1	·35168	5	1	·55366		45
	30	3	1	·61990	5	1	·39009		30
	45	3	1	·88740	5	1	·22536		15
32		3	2	·15419	5	1	·05946	58	
	15	3	2	·42024	5	0	·89240		45
	30	3	2	·68557	5	0	·72418		30
	45	3	2	·95016	5	0	·55481		15
33		3	3	·21401	5	0	·38428	57	
	15	3	3	·47711	5	0	··21261		45
	30	3	3	·73946	5	0	·03978		30
	45	3	4	·00105	·4	11	·86581		15
34		3	4	·26189	4	11	·69071	56	
	15	3	4	·52195	4	11	·51446		45
	30	3	4	·78125	4	11	·33709		30
	45	3	5	·03977	4	11	·15858		15
35		3	5	·29750	4	10	·97894	55	
	15	3	5	·55445	4	10	·79819		45
	30	3	5	·81062	4	10	·61632		30
	45	3	6	·06598	4	10	·43333		15
36		3	6	·32054	4	10	·24922	54	
	15	3	6	·57429	4	10	·06401		45
	30	3	6	·82724	4	9	·87770		30
	45	3	7	·07937	4	9	·69027		15
37		3	7	·33068	4	9	·50176	53	
	15	3	7	·58117	4	9	·31214		45
	30	3	7	·83082	4	9	·12144		30
	45	3	8	·07965	4	8	·92965		15

PERPENDICULAR	BASE	ANGLE

| ANGLE | | BASE | | | PERPENDICULAR | | | | |
Deg.	Min.	Feet	Ins.	Decimals	Feet	Ins.	Decimals	Deg.	Min.
38		3	8	·32763	4	8	·73678	52	
	15	3	8	·57476	4	8	·54282		45
	30	3	8	·82105	4	8	·34779		30
	45	3	9	·06649	4	8	·15168		15
39		3	9	·31107	4	7	·95451	51	
	15	3	9	·55478	4	7	·75627		45
	30	3	9	·79763	4	7	·55697		30
	45	3	10	·03961	4	7	·35661		15
40		3	10	·28071	4	7	·15520	50	
	15	3	10	·52093	4	6	·95274		45
	30	3	10	·76026	4	6	·74923		30
	45	3	10	·99871	4	6	·54468		15
41		3	11	·23625	4	6	·33909	49	
	15	3	11	·47290	4	6	·13246		45
	30	3	11	·70864	4	5	·92481		30
	45	3	11	·94348	4	5	·71613		15
42		4	0	·17740	4	5	·50643	48	
	15	4	0	·41041	4	5	·29570		45
	30	4	0	·64250	4	5	·08396		30
	45	4	0	·87365	4	4	·87122		15
43		4	1	·10388	4	4	·65747	47	
	15	4	1	·33318	4	4	·44271		45
	30	4	1	·56153	4	4	·22696		30
	45	4	1	·78894	4	4	·01021		15
44		4	2	·01540	4	3	·79247	46	
	15	4	2	·24092	4	3	·57374		45
	30	4	2	·46547	4	3	·35403		30
	45	4	2	·68906	4	3	·13335		15
45		4	2	·91169	4	2	·91169	45	
PERPENDICULAR					BASE			ANGLE	

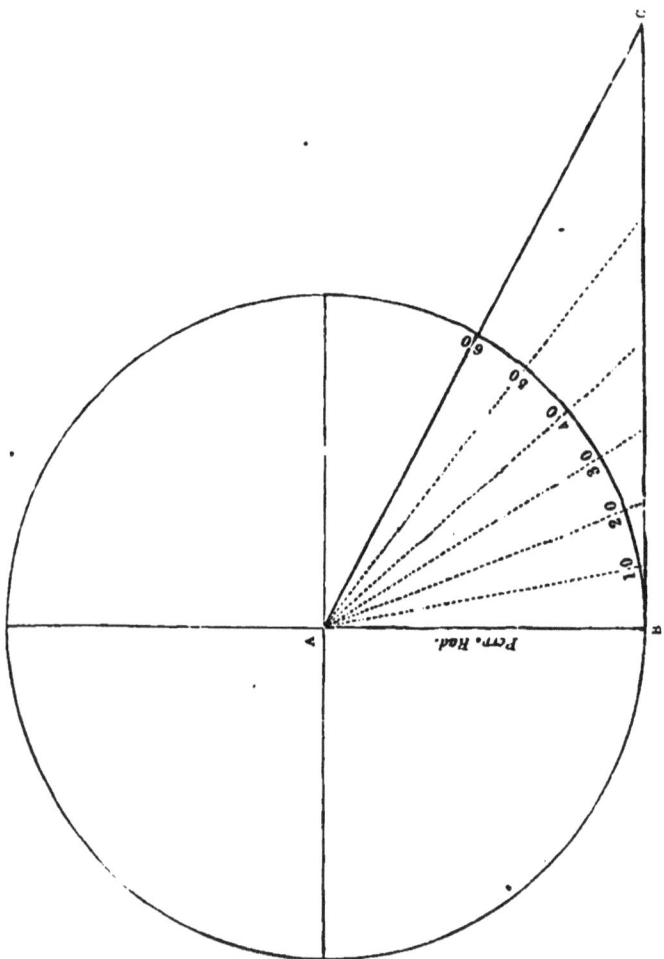

SECOND TABLE.

PERPENDICULAR RADIUS,

ONE FATHOM.

ANGLE		BASE				HYPOTHENUSE			
Deg.	Min.	Fath.	Feet	Ins.	Decimals	Fath.	Feet	Ins.	Decimals
1		0	0	1	·2568	1	0	0	·0108
	15	0	0	1	·5710	1	0	0	·0171
	30	0	0	1	·8854	1	0	0	·0247
	45	0	0	2	·1998	1	0	0	·0335
2		0	0	2	·5143	1	0	0	·0432
	15	0	0	2	·8289	1	0	0	·0554
	30	0	0	3	·1435	1	0	0	·0684
	45	0	0	3	·4582	1	0	0	·0828
3		0	0	3	·7728	1	0	0	·0986
	15	0	0	4	·0882	1	0	0	·1159
	30	0	0	4	·4035	1	0	0	·1346
	45	0	0	4	·7189	1	0	0	·1544
4		0	0	5	·0328	1	0	0	·1757
	15	0	0	5	·3496	1	0	0	·1980
	30	0	0	5	·6664	1	0	0	·2171
	45	0	0	5	·9825	1	0	0	·2484
5		0	0	6	·2993	1	0	0	·2736
	15	0	0	6	·6168	1	0	0	·3024
	30	0	0	6	·9336	1	0	0	·3312
	45	0	0	7	·2497	1	0	0	·3636
6		0	0	7	·5672	1	0	0	·3960
	15	0	0	7	·8841	1	0	0	·4298
	30	0	0	8	·2008	1	0	0	·4658
	45	0	0	8	·5212	1	0	0	·5026
7		0	0	8	·8402	1	0	0	·5400
	15	0	0	9	·1584	1	0	0	·5803
	30	0	0	9	·4788	1	0	0	·6192
	45	0	0	9	·7992	1	0	0	·6624
8		0	0	10	·1189	1	0	0	·7056
	15	0	0	10	·4393	1	0	0	·7531
	30	0	0	10	·7604	1	0	0	·7992
	45	0	0	11	·0808	1	0	0	·8474

ANGLE		BASE				HYPOTHENUSE			
Deg.	Min.	Fath.	Feet	Ins.	Decimals	Fath.	Feet	Ins.	Decimals
9		0	0	11	·4034	1	0	0	·8971
	15	0	0	11	·7259	1	0	0	·9482
	30	0	1	0	·0485	1	0	1	·0008
	45	0	1	0	·3696	1	0	1	·0548
10		0	1	0	·6936	1	0	1	·1088
	15	0	1	1	·0176	1	0	1	·1664
	30	0	1	1	·3416	1	0	1	·2262
	45	0	1	1	·6692	1	0	1	·2859
11		0	1	1	·9954	1	0	1	·3476
	15	0	1	2	·3215	1	0	1	·4106
	30	0	1	2	·6484	1	0	1	·4750
	45	0	1	2	·9760	1	0	1	·5410
12		0	1	3	·3036	1	0	1	·6085
	15	0	1	3	·6326	1	0	1	·6775
	30	0	1	3	·9617	1	0	1	·7481
	45	0	1	4	·2914	1	0	1	·8202
13		0	1	4	·6219	1	0	1	·8939
	15	0	1	4	·9538	1	0	1	·9691
	30	0	1	5	·2857	1	0	2	·0459
	45	0	1	5	·6178	1	0	2	·1242
14		0	1	5	·9496	1	0	2	·2042
	15	0	1	6	·2858	1	0	2	·2857
	30	0	1	6	·6192	1	0	2	·3688
	45	0	1	6	·9576	1	0	2	·4535
15		0	1	7	·2888	1	0	2	·5399
	15	0	1	7	·6294	1	0	2	·6338
	30	0	1	7	·9670	1	0	2	·7174
	45	0	1	8	·3062	1	0	2	·8087
16		0	1	8	·6456	1	0	2	·9015
	15	0	1	8	·9858	1	0	2	·9961
	30	0	1	9	·3271	1	0	3	·0443
	45	0	1	9	·6695	1	0	3	·1902
17		0	1	10	·0126	1	0	3	·2898
	15	0	1	10	·3560	1	0	3	·3911
	30	0	1	10	·7003	1	0	3	·4941
	45	0	1	11	·0472	1	0	3	·5988
18		0	1	11	·3942	1	0	3	·7053
	15	0	1	11	·6220	1	0	3	·8135
	30	0	2	0	·0905	1	0	3	·9234
	45	0	2	0	·4204	1	0	4	·0352

ANGLE		BASE				HYPOTHENUSE			
Deg.	Min.	Fath.	Feet	Ins.	Decimals	Fath.	Feet	Ins.	Decimals
19		0	2	0	·7916	1	0	4	·1487
	15	0	2	1	·1435	1	0	4	·2640
	30	0	2	1	·4965	1	0	4	·3811
	45	0	2	1	·8506	1	0	4	·5000
20		0	2	2	·2058	1	0	4	·6208
	15	0	2	2	·5622	1	0	4	·7434
	30	0	2	2	·9197	1	0	4	·8679
	45	0	2	3	·2783	1	0	4	·9942
21		0	2	3	·6262	1	0	5	·1224
	15	0	2	3	·9993	1	0	5	·2526
	30	0	2	4	·3615	1	0	5	·3846
	45	0	2	4	·7251	1	0	5	·5186
22		0	2	5	·0899	1	0	5	·6545
	15	0	2	5	·4560	1	0	5	·7924
	30	0	2	5	·8234	1	0	5	·9323
	45	0	2	6	·1921	1	0	6	·0741
23		0	2	6	·5622	1	0	6	·2179
	15	0	2	6	·9336	1	0	6	·3638
	30	0	2	7	·3065	1	0	6	·5117
	45	0	2	7	·6807	1	0	6	·6617
24		0	2	8	·0565	1	0	6	·8138
	15	0	2	8	·4336	1	0	6	·9679
	30	0	2	8	·8123	1	0	7	·1242
	45	0	2	9	·1924	1	0	7	·2826
25		0	2	9	·5741	1	0	7	·4432
	15	0	2	9	·9574	1	0	7	·6059
	30	0	2	10	·3422	1	0	7	·7708
	45	0	2	10	·7287	1	0	7	·9380
26		0	2	11	·1167	1	0	8	·1073
	15	0	2	11	·5065	1	0	8	·2789
	30	0	2	11	·8979	1	0	8	·4528
	45	0	3	0	·2910	1	0	8	·6290
27		0	3	0	·6858	1	0	8	·8075
	15	0	3	1	·0824	1	0	8	·9883
	30	0	3	1	·4808	1	0	9	·1714
	45	0	3	1	·8810	1	0	9	·3571
28		0	3	2	·2831	1	0	9	·5450
	15	0	3	2	·6870	1	0	9	·7354
	30	0	3	3	·0928	1	0	9	·9283
	45	0	3	3	·5005	1	0	10	·1232

ANGLE		BASE				HYPOTHENUSE			
Deg.	Min.	Fath.	Feet	Ins.	Decimals	Fath.	Feet	Ins.	Decimals
29		0	3	3	·9102	1	0	10	·3212
	15	0	3	4	·3219	1	0	10	·5220
	30	0	3	4	·7356	1	0	10	·7251
	45	0	3	5	·1514	1	0	10	·9303
30		0	3	5	·5692	1	0	11	·1384
	15	0	3	5	·9892	1	0	11	·3494
	30	0	3	6	·4112	1	0	11	·5625
	45	0	3	6	·8355	1	0	11	·7788
31		0	3	7	·2620	1	0	11	·9976
	15	0	3	7	·6907	1	1	0	·2192
	30	0	3	8	·1216	1	1	0	·4436
	45	0	3	8	·5550	1	1	0	·7008
32		0	3	8	·9906	1	1	0	·9008
	15	0	3	9	·4286	1	1	1	·1338
	30	0	3	9	·8691	1	1	1	·3696
	45	0	3	10	·3119	1	1	1	·6084
33		0	3	10	·7573	1	1	1	·8501
	15	0	3	11	·2053	1	1	2	·0949
	30	0	3	11	·6558	1	1	2	·3427
	45	0	4	0	·1088	1	1	2	·5937
34		0	4	0	·5646	1	1	2	·8477
	15	0	4	0	·9931	1	1	3	·1049
	30	0	4	1	·4842	1	1	3	·3653
	45	0	4	1	·9482	1	1	3	·6289
35		0	4	2	·4149	1	1	3	·8958
	15	0·	4	2	·8846	1	1	4	·1660
	30	0	4	3	·3571	1	1	4	·4395
	45	0	4	3	·8326	1	1	4	·7165
36·		0	4	4	·3111	1	1	4	·9969
	15	0	4	4	·7914	1	1	5	·2808
	30	0	4	5	·2772	1	1	5	·6819
	45	0	4	5	·5650	1	1	5	·8591
37		0	4	6	·2559	1	1	6	·1538
	15	0	4	6	·7501	1	1	6	·4520
	30	0	4	7	·2475	1	1	6	·7540
	45	0	4	7	·7483	1	1	7	·0597
38		0	4	8	·2526	1	1	7	·3693
	15	0	4	8	·7602	1	1	7	·6831
	30	0	4	9	·2834	1	1	8	·0000
	·45	0	4	9	·7861	1	1	8	·3214

ANGLE		BASE.				HYPOTHENUSE			
Deg.	Min.	Fath.	Feet	Ins.	Decimals	Fath.	Feet	Ins.	Decimals
39		0	4	10	·3044	1	1	8	·6467
	15	0	4	10	·8265	1	1	8	·9761
	30	0	4	11	·3522	1	1	9	·3096
	45	0	4	11	·8818	1	1	9	·6473
40		0	5	0	·4152	1	1	9	·9893
	15	0	5	0	·9525	1	1	10	·0956
	30	0	5	1	·4938	1	1	10	·6863
	45	0	5	2	·0392	1	1	11	·0413
41		0	5	2	·5886	1	1	11	·4009
	15	0	5	3	·1420	1	1	11	·7651
	30	0	5	3	·7002	1	2	0	·1338
	45	0	5	4	·2624	1	2	0	·5049
42		0	5	4	·8291	1	2	0	·8855
	15	0	5	5	·4002	1	2	1	·2686
	30	0	5	5	·9758	1	2	1	·6566
	45	0	5	6	·5705	1	2	2	·0496
43		0	5	7	·1411	1	2	2	·4476
	15	0	5	7	·7308	1	2	2	·8507
	30	0	5	8	·3254	1	2	3	·2591
	45	0	5	8	·9250	1	2	3	·6727
44		0	5	9	·5296	1	2	4	·0918
	15	0	5	10	·1393	1	2	4	·5163
	30	0	5	10	·7542	1	2	4	·9463
	45	0	5	11	·3744	1	2	5	·3820
45		1	0	0	·0000	1	2	5	·8234
	15	1	0	0	·6311	1	2	6	·2706
	30	1	0	1	·2677	1	2	6	·7237
	45	1	0	1	·9101	1	2	7	·1828
46		1	0	2	·5852	1	2	7	·6481
	15	1	0	3	·2122	1	2	8	·1195
	30	1	0	3	·8722	1	2	8	·5973
	45	1	0	4	·5382	1	2	9	·0814
47		1	0	5	·2105	1	2	9	·5721
	15	1	0	5	·8892	1	2	10	·0694
	30	1	0	6	·5742	1	2	10	·5735
	45	1	0	7	·2658	1	2	11	·0844
48		1	0	7	·9641	1	2	11	·6023
	15	1	0	8	·6692	1	3	0	·1273
	30	1	0	9	·3812	1	3	0	·6596
	45	1	0	10	·1003	1	3	1	·1991

ANGLE		BASE				HYPOTHENUSE			
Deg.	Min.	Fath.	Feet	Ins.	Decimals	Fath.	Feet	Ins.	Decimals
49		1	0	10	·8265	1	3	1	·7462
	15	1	0	11	·5601	1	3	2	·3009
	30	1	1	0	·3012	1	3	2	·8634
	45	1	1	1	·0498	1	3	3	·4337
50		1	1	1	·8062	1	3	4	·0122
	15	1	1	2	·5699	1	3	4	·5987
	30	1	1	3	·3430	1	3	5	·1936
	45	1	1	4	·1236	1	3	5	·7970
51		1	1	4	·9126	1	3	6	·4091
	15	1	1	5	·7101	1	3	7	·0300
	30	1	1	6	·5164	1	3	7	·6599
	45	1	1	7	·3316	1	3	8	·2990
52		1	1	8	·1560	1	3	8	·9474
	15	1	1	8	·9893	1	3	9	·6053
	30	1	1	9	·8322	1	3	10	·2729
	45	1	1	10	·6848	1	3	10	·9505
53		1	1	11	·5472	1	3	11	·6381
	15	1	2	0	·4197	1	4	0	·3360
	30	1	2	1	·3024	1	4	1	·0445
	45	1	2	2	·1956	1	4	1	·7636
54		1	2	3	·0995	1	4	2	·4937
	15	1	2	4	·0143	1	4	3	·2350
	30	1	2	4	·9403	1	4	3	·9876
	45	1	2	5	·8776	1	4	4	·7520
55		1	2	6	·8267	1	4	5	·5282
	15	1	2	7	·7876	1	4	6	·3165
	30	1	2	8	·7606	1	4	7	·1172
	45	1	2	9	·7458	1	4	7	·9306
56		1	2	10	·7444	1	4	8	·7570
	15	1	2	11	·7556	1	4	9	·6966
	33	1	3	0	·7801	1	4	10	·4497
	45	1	3	1	·8182	1	4	11	·3165
57		1	3	2	·8703	1	5	0	·1976
	15	1	3	3	·9365	1	5	1	·0932
	30	1	3	5	·0174	1	5	2	·0034
	45	1	3	6	·1131	1	5	2	·9287
58		1	3	7	·2241	1	5	3	·8697
	15	1	3	8	·3507	1	5	4	·8265
	30	1	3	9	·4933	1	5	5	·7994
	45	1	3	10	·6523	1	5	6	·7890

ANGLE		BASE				HYPOTHENUSE			
Deg.	Min.	Fath.	Feet	Ins.	Decimals	Fath.	Feet	Ins.	Decimals
59		1	3	11	·7281	1	5	7	·7955
	15	1	4	1	·0211	1	5	8	·8194
	30	1	4	2	·2317	1	5	9	·8612
	45	1	4	3	·4604	1	5	10	·9212
60		1	4	4	·7077	2	0	0	·0000

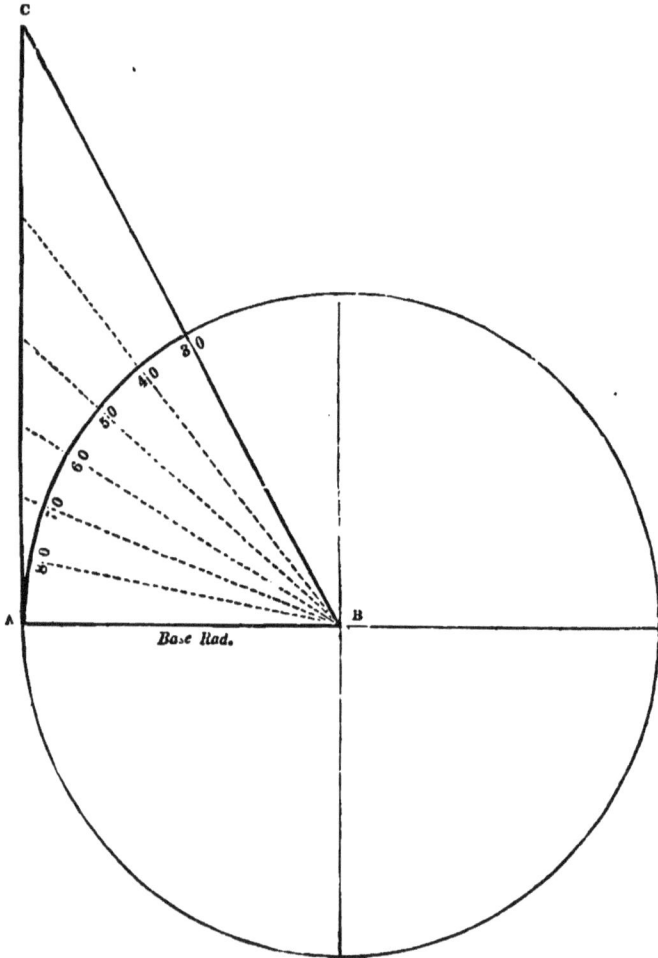

Base Rad.

THIRD TABLE.

BASE RADIUS,

ONE FATHOM.

ANGLE	HYPOTHENUSE				PERPENDICULAR			
Degrees	Fath.	Feet	Ins.	Decimals	Fath.	Feet	Ins.	Decimals
1	57	1	9	·50554	57	1	8	·87726
2	28	3	11	·06698	28	3	9	·81022
3	19	0	7	·72726	19	0	5	·84186
4	14	2	0	·16226	14	1	9	·64795
5	11	2	10	·10734	11	2	6	·96374
6	9	3	4	·80760	9	3	1	·03424
7	8	1	2	·79665	8	0	10	·39294
8	7	1	1	·34135	7	0	8	·30662
9	6	2	4	·25663	6	1	10	·59011
10	5	4	6	·63148	5	4	0	·33229
11	5	1	5	·34070	5	0	10	·40789
12	4	4	10	·30087	4	4	2	·73337
13	4	2	8	·06963	4	1	11	·86626
14	4	0	9	·61672	4	0	0	·77622
15	3	5	2	·18664	3	4	4	·70766
16	3	3	9	·21278	3	2	11	·09384
17	3	2	6	·26186	3	1	7	·50139
18	3	1	4	·99690	3	0	5	·59321
19	3	0	5	·15185	2	5	5	·10318
20	2	5	6	·51392	2	4	5	·81837
21	2	4	8	·91082	2	3	7	·56641
22	2	4	0	·20164	2	2	10	·20626
23	2	3	4	·26994	2	2	1	·62137
24	2	2	9	·01872	2	1	5	·71465
25	2	2	2	·86651	2	0	10	·40450
26	2	1	8	·24438	2	0	3	·62187
27	2	1	2	·59363	1	5	9	·30796
28	2	0	9	·36392	1	5	3	·41231
29	2	0	4	·51190	1	4	9	·89144
30	2	0	0	·00000	1	4	4	·70766
31	1	5	7	·79549	1	3	11	·82812
32	1	5	3	·86975	1	3	7	·22408

ANGLE	HYPOTHENUSE				PERPENDICULAR			
Degrees	Fath.	Feet	Ins.	Decimals	Fath.	Feet	Ins.	Decimals
33	1	5	0	·19765	1	3	2	·87028
34	1	4	8	·75699	1	2	10	·74439
35	1	4	5	·52817	1	2	6	·82666
36	1	4	2	·49371	1	2	3	·09950
37	1	3	11	·63809	1	1	11	·54722
38	1	3	8	·94738	1	1	8	·15579
39	1	3	6	·40913	1	1	4	·91260
40	1	3	4	·01211	1	1	1	·80626
41	1	3	1	·74622	1	0	10	·82652
42	1	2	11	·60281	1	0	7	·96410
43	1	2	9	·57201	1	0	5	·21055
44	1	2	7	·64807	1	0	2	·55818
45	1	2	5	·82338	1	0	0	·00000
46	1	2	4	·09178	0	5	9	·52959
47	1	2	2	·44758	0	5	7	·14108
48	1	2	0	·88555	0	5	4	·82909
49	1	1	11	·40094	0	5	2	·58864
50	1	1	9	·98932	0	5	0	·41517
51	1	1	8	·64669	0	4	10	·30445
52	1	1	7	·36931	0	4	8	·25256
53	1	1	6	·15377	0	4	6	·25589
54	1	1	4	·99689	0	4	4	·31106
55	1	1	3	·89577	0	4	2	·41494
56	1	1	2	·84768	0	4	0	·56461
57	1	1	1	·85016	0	3	10	·75735
58	1	1	0	·90084	0	3	8	·99060
59	1	0	11	·99760	0	3	7	·26196
60	1	0	11	·13844	0	3	5	·56922

LEVELLING.

RULE. — Add all the perpendiculars together. for the base line or horizontal distance, and subtract the bases made by the angles of elevation and depression one from the other, for the perpendicular or difference of height.*

EXAMPLE.

Being required to level an irregular piece of ground, I measured in a south-west direction 64 yards from A to B, on an angle of depression 9° 45′; from this

---

* The altitudes of irregular hills are generally ascertained by the assistance of a spirit level and perpendicular poles, and if the ground rise and descend alternately, the differences between the heights of the poles are added when ascending, and subtracted when descending, in order to determine the different elevations and depressions of the ground: the foregoing rule and method will be found far more correct and masterly, remembering always that the height of the instrument be accounted for, which may easily be done by taking the observation from a staff or target the same height as the instrument.

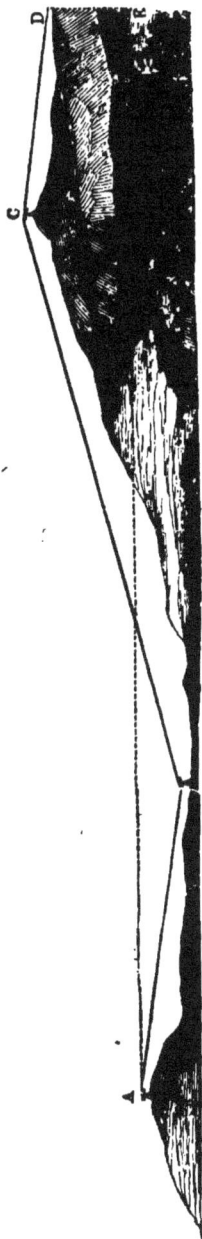

station I measured 120 yards from B to C, in the same cardinal direction, on an angle of elevation 16° 30′; and from thence to the extent of the ground the line on the same course measured 44 yards from C to D, and the angle of depression 7°: required the base line or horizontal distance from the place where the levelling was begun, to the point where it was ended; also, how much higher or lower the ground is at the place where the operation terminated, than where it commenced.

PERENDICULARS.

	ft.	in.		fath.		ft.	in.
∠ 9° 45′ =	5	10·96004	×	32	=	189	2·72128
∠ 16° 30′ =	5	9·03502	×	60	=	345	2·10120
∠ 7° 0′ =	5	11·46333	×	22	=	131	0·19326
						3)665	5·01574
					A E	221 yds.	2 ft. 5 in.

BASES.

		ft.	in.		fath.		ft.	in.
Depression.	∠ 9° 45′ =	1	0·19316	×	32	=	32	6·18112
	∠ 7° 0′ =	0	8·77459	×	22	=	18	1·04098
							50	7·22210
Elevation.	∠ 16° 30′ =	1	8·44910	×	60	=	102	2·94600
							3) 51	7·72390
						E D	17 yds.	0 ft. 7½ in.

ANSWER.

	yds.	ft.	in.
Horizontal distance A E	221	2	5
Elevation E D	17	0	7½

HORIZONTAL

OR

TRAVERSE DIALLING.

PLANE sailing in navigation and horizontal dialling in mining, are nothing more than the practice of right-angled trigonometry, calling the hypothenuse the distance, the perpendicular the difference of latitude, the base the departure, and the angle opposite the base the course; consequently any range of dialling, however complicated and extensive, may be reduced into a single triangle, the perpendicular of which will either be the east and west, or north and south line, according to the main direction or bearing of the work; the hypothenuse will be the actual length of the dialling in a right line from the point of setting out to the termination; the base will be the distance the terminating point will fall right or left of the perpendicular; and the angle made by the hypothenuse with the perpendicular will be the final course or direction of the work.

It therefore follows, that the general practice of repeating or retracing a course of underground dialling on the surface may be avoided, and thereby the difficulties and dangers arising from obstructions, irregular ground, and the attraction of the magnet by iron, which always abounds in the vicinity of a mine, be done away.

What is said of Mercator's sailing may, in the chief respect, be applied to horizontal dialling, viz. : ' It is the art of finding on a plane surface the motion of a ship upon any assigned course by the compass, which shall be true in latitude, longitude, and distance sailed ;' and certainly this includes the whole theory and practice of navigation ; and if any method could be devised for measuring a ship's course and distance truly, nothing would be wanting:—also in dialling, it is only required to find a method for reducing the various windings and angles of a level or adit into a right line, and discovering the real extent and direction of that line, to complete the art.

But not to occupy the reader's time in telling him what he well knows already, we shall proceed to introduce the process for obtaining the length and bearing of a course of traverse dialling by the trigonometrical tables.

The first thing to be attended to is the statement of the work, or so placing the drafts that there may be no confusion in the operation, and that the perpendiculars and bases may fall on their proper sides.

In order to succeed in this essential matter, which may be considered the foundation of the work, note on which cardinal point the main direction of your dialling runs, whether east, west, north, or south, and reckon off your degrees right or left from that line: thus—if your dialling runs easterly or westerly, let the equator, or east and west line, be the point for numbering off your angles — if northerly or southerly, the meridian or north and south line; consequently this line will be the perpendicular of every triangle in the operation that comes within the sweep of half the circle, or 180°; and should any of the drafts return beyond the north or south points, or exceed 90° right or left of the east point, then the angle must be counted from the west towards the north or south, as the draft may happen to incline.

' This being done it is evident that on a course of east and west dialling, the bases north and bases south must be subtracted one from the other, and the remainder will be the departure or base line north or south as the dialling may have prevailed on this or that side, and if any of the drafts have gone westerly, then the perpendiculars west must be subtracted from the perpendiculars east, for the real length of the perpendiculars; but if the dialling has prevailed most in a westerly direction, the perpendicular will lie on that side : in short, as· a matter of course, either for the difference of latitude, or rather difference of longitude in this case (the perpendicular), or for the departure (the base), the less number must be taken from the greater, and the differences will show the sides on which the operation lies.

This process must all be performed by the first table, where the hypothenuse is given, because in every case the actual measured line will be the longest side of the triangle, and after stating the work, as before directed, take out the numbers standing against the given angles in the table and multiply them respectively by the length of the hypothenuse, reduced into fathoms and parts (if any), and place them in their proper positions until the whole has been calculated; then take the sum of the bases north and south one from the other, and the sum of the perpendiculars east and west one from the other; the perpendicular remainders will show the east and west line, and the bases the distance the dialling has extended north or south of that line.

The work is now brought to that case where the difference of latitude and departure is given to find the course and distance, and in order to avoid the necessity of introducing extensive and intricate tables, used by navigators for this purpose, we shall have recourse to one simple act of instrumental operation, and as two sides of the triangle are given, the thing may be quickly and safely performed; thus—draw the base the given length by a scale of equal parts, raise the perpendicular on one end of the base (and of course at right angles therewith), and mark off the given length, draw the hypothenuse, and the triangle will be complete: then, by the same scale, measure the hypothenuse, and it will be the actual length of the dialling in a right line, from beginning to end; then, with a protractor or scale of chords, measure the angle opposite the departure or base, and it will be the true course, bearing, or direction of the extreme points.

The degrees on the miner's compass are generally graduated from 1 to 360, and are figured toward the left hand, consequently 90° stands at the west point, 180° at south, 270° at the east, and ends with 360° at the north; and when the same course is to be pursued, that is, when the angles are to be taken and the drafts measured again, there will be no necessity for finding the real direction of the line, for as the sights are always fixed, the dialler need only be careful to observe that the needle stands at the same degree as in the original course: but when the operation is to be plotted or trigonometrically proved, there will be a necessity for ascertaining the actual bearing of every draft in the work, and this may be done by the following rule.

RULE.

(SIGHTS FIXED NORTH AND SOUTH.)

When the needle rests on any degree.		The direction of the sights or the course of the dialling will be	
	From 1 to 90 N. to W.		E. of N. Comp. N. of E.
	From 90 to 180 W. to S.		S. of E. Comp. E. of S.
	From 180 to 270 S. to E.		W. of S. Comp. S. of W.
	From 270 to 360 E. to N.		N. of W. Comp. W. of N.

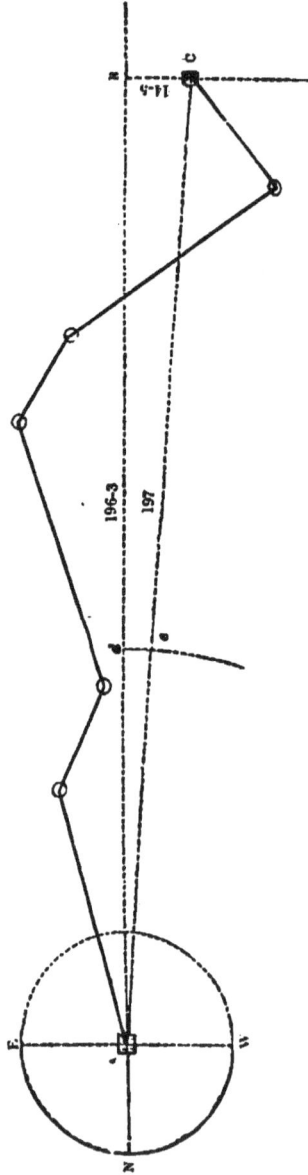

Scale 40 feet to an inch.

EXAMPLE 1.

It is required to sink a perpendicular shaft on the end of a level whose angles and drafts measured as follows, viz. :

		ft.	in.	fath.	ft.	in.
∠16°	30′ E. of S.	53	6 or	8	5	6
∠26°	0′ W. of S.	22	11 or	3	4	11
∠19°	0′ E. of S.	58	0 or	9	4	0
∠34°	30′ W. of S.	21	6 or	3	3	6
∠57°	30′ W. of S.	53	8 or	8	5	8
∠39°	30′ E. of S.	29	10 or	4	5	10

What distance is the end C (in the annexed plate), where the dialling was finished, from the engine shaft A, where the dialling was begun, and what is the bearing of the line A C, or how many degrees are contained in the angle B A C?

OPERATION.

BASES.

		ft.	in.		fath.	ft.	in.		ft.	in.
E. of S. 16°½	= 1	8·44910		×	8	5	6	= 15	2·33790	
W. of S. 26°	= 2	7·56272		×	3	4	11	= 10	0·53916	
E. of S. 19°	= 1	11·44091		×	9	4	0	= 18	10·58864	
W. of S. 34°½	= 3	4·78125		×	3	3	6	= 12	2·13235	
W. of S. 57°½	= 5	0·72418		×	8	5	8	= 45	3·18762	
E. of S. 39°¾	= 3	9·79763		×	4	5	10	= 18	11·55815	

	ft.	in.
Sum of bases W. of S.	67	5·85913
Sum of bases E. of S.	53	0·48469
Base or departure Westerly B C=	14	5·37444

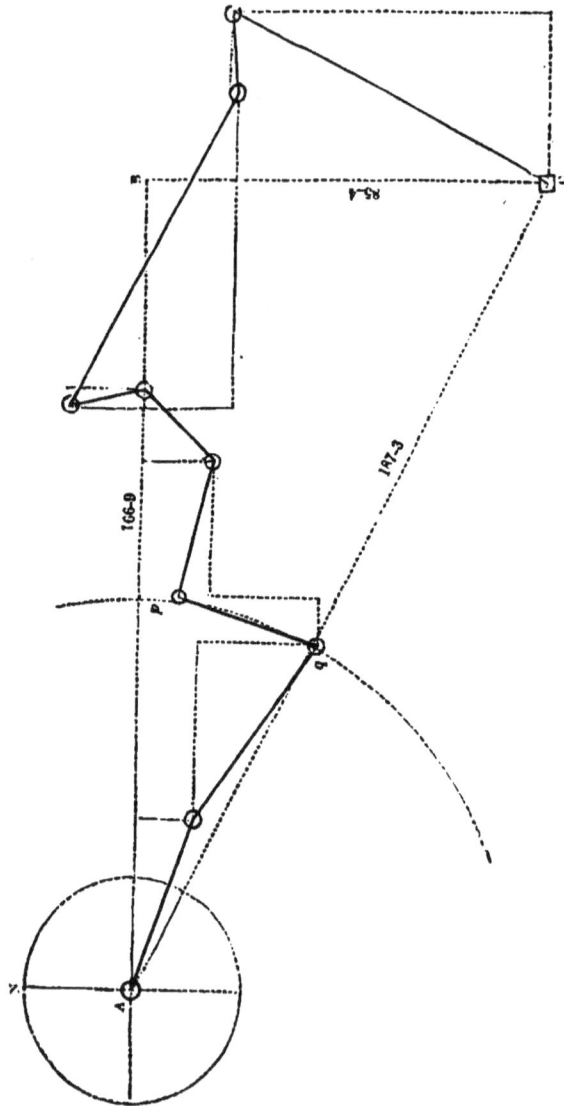

PERPENDICULARS.

ft.	in.	fath.	ft.	in.	ft.	in.
$\angle\,16^\circ\frac{1}{2} = 5$	9·03502	× 8	5	6 = 51		3·56228
$\angle\,26^\circ = 5$	4·71317	× 3	4	11 = 20		7·16851
$\angle\,19^\circ = 5$	8·07734	× 9	4	0 = 54		10·06606
$\angle\,34^\circ\frac{1}{2} = 4$	11·33709	× 3	3	6 = 17		8·62417
$\angle\,57^\circ\frac{1}{2} = 3$	2·68557	× 8	5	8 = 28		10·05018
$\angle\,39^\circ\frac{1}{2} = 4$	7·55697	× 4	5	10 = 23		0·04485

Perpendicular or diff. of latitude, A B   196   3·51605

THEN—BY CONSTRUCTION.

Draw two lines at right angles, as A B and B C, and of an indefinite length, take 196 feet 3½ inches in your compasses from a scale of equal parts, and with one foot in the right angle B, point off the distance B A for the perpendicular. Again take 14 feet 5¼ inches from the same scale, and apply it to the other line B C for the base; draw the hypothenuse to join A C, which by the same scale will be found to measure 197 feet.

FOR THE ANGLE.

With the chord of 60° in your compasses and centre A, describe an arc $e\,d$ cutting A B and A C in d and e; then take the distance $e\,d$ in your compasses, and setting one foot on the brass pin at the beginning of the chords on your scale, observe how many degrees the other foot reaches to, which will be 4° 15′ for the arc $e\,d$ or angle B A C.

ANSWER.

197 feet, on an angle of 4° 15′ west of south.

EXAMPLE 2.`

Given the following course of traverse dialling, viz. :

		ft.	in.	fath.	ft.	in.
162°	= ∠ 18° 0′ S. of E.	36	0 or	6	0	0
143°¾	= ∠ 36° 15′ S. of E.	44	4 or	7	2	4
16°½	= ∠ 73° 30′ N. of E.	30	9 or	5	0	9
257°¼	= ∠ 12° 45′ S. of E.	28	6 or	4	4	6
45°	= ∠ 45° 0′ N. of E.	17	10 or	2	5	10
7°¾	= ∠ 82° 15′ N. of W.	15	3 or	2	3	3
152°¼	= ∠ 27° 30′ S. of E.	72	0 or	12	0	0
87°½	= ∠ 2° 30′ N. of E.	16	0 or	2	4	0
204°½	= ∠ 65° 30′ S. of W.	73	0 or	12	1	0

(Degrees at which the needle stood on the outer circle of the dial.)

Required the distance and bearing of the extreme points A C.

OPERATION.

BASES SOUTHERLY.

	ft.	in.	fath.	ft.	in.	ft.	in.
18° 0′ = 1	10·24922	× 6	0	0 = 11	1·49522
36° 15′ = 3	6·57429	× 7	2	4 = 26	2·57669
12° 45′ = 1	3·89021	× 4	4	6 = 6	3·44849
27° 30′ = 2	9·24590	× 12	0	0 = 33	2·95080
65° 30′ = 5	5·51721	× 12	1	0 = 66	5·12605

143   3·59725

BASES NORTHERLY.

	ft.	in.	fath.	ft.	in.	ft.	in.
73° 30′ = 5	9·03502	× 5	0	9 = 29	5·80448
45° 0′ = 4	2·91169	× 2	5	10 = 12	7·14507
82° 15′ = 5	11·34234	× 2	3	3 = 15	1·33685
2° 30′ = 0	3·14060	× 2	4	0 =	0	8·37150

57   10·65790

	ft.	in.
From bases southerly	= 143	3·59725
Subtract bases northerly	= 57	10·65790
Departure B C	85	4·93935

PERPENDICULARS EASTERLY.

	ft.	in.	fath.	ft.	in.	ft.	in.
18° 0′ =	5	8·47607	× 6	0	0 =	34	2·85642
36° 15′ =	4	10·06401	× 7	2	4 =	35	9·02851
73° 30′ =	1	8·44910	× 5	0	9 =	8	10·80163
12° 45′ =	5	10·22464	× 4	4	6 =	27	9·56704
45° 0′ =	4	2·91169	× 2	5	10 =	12	7·14507
27° 30′ =	5	3·86478	×12	0	0 =	63	10·37736
2° 30′ =	5	11·93147	× 2	4	0 =	15	11·81724
						199	1·59327

PERPENDICULARS WESTERLY.

	ft.	in.	fath.	ft.	in.	ft.	in.
82° 15′ =	0	9·70926	× 2	3	3 =	2	0·67815
65° 30′ =	2	5·85791	×12	1	0 =	30	3·27124
						32	3·94939

	ft.	in.
From perpendiculars east	= 199	1·59327
Subtract perpendiculars west	= 32	3·94939
A B	166	9·64388

	ft.	in.
Perp. or east and west line. A B	166	9·64388
Base south of east C B	85	4·93935

Then by construction (as before) the hypothenuse A C will be found 187 feet 3 inches, and the angle *p q* 27 degrees, south of east.

THE

PRACTICAL MINER'S GUIDE.

PART II.

——◆——

INTRODUCTION.

THE qualifications necessary to constitute an accomplished miner are more numerous and difficult of attainment than is generally imagined, even
by persons deeply interested in mining affairs ; and
although it may not be expected that everyone
who fills a mining situation should be an adept in
all the various branches of the art, yet it is certainly highly desirable that agents, who have the
management of large adventures, should possess a
general knowledge of everything connected with
the profession of a miner.

The following miscellaneous subjects are essential to the practical miner, and require no comment
to set forth their utility ; they may also be found
useful and interesting to persons not immediately
engaged in mining pursuits.

The first article consists in a description of the
art of assaying silver ; and as this has hitherto
been a secret in the possession of but few persons,

it is expected that it will form an acceptable part of the work, especially as it will come abroad at a time when foreign mining speculations (where the seat of action is principally among the precious metals) abound beyond all precedent.

The next part of the work contains a plain statement of the method of assaying copper, including the established process of one of the most experienced and respectable copper-assayers in the county of Cornwall.—Rules for assaying lead and tin follow in succession, and this part of the treatise concludes with a description of the manner of extracting silver from copper ore, or of discovering the quantity of silver it contains ; and probably this article also may be productive of beneficial effects to the mining interest, as there is great reason to believe that a considerable proportion of silver is contained in the ores produced from many of our copper mines. The method is very simple, and the trial may be quickly and satisfactorily made.

The subsequent part of the work is described in the table of contents.

ASSAY OF SILVER ORE.

—•—

Sample — 1 ounce avoirdupois, pulverised and sifted through a fine hair sieve, then well mixed in the scoop with the following flux, viz. :

Red lead* . . . . 2 oz.

Red tartar . . . . 5 dwts.

Nitre . . . . 9 dwts.

Borax . . . : 4 dwts.

Lime . . . . $\frac{1}{4}$ oz.

Salt . . . . 2 oz.

Fluor spar (bruised) . . $\frac{1}{4}$ oz.

Smelt the ore in a wrought-iron crucible; if this cannot be conveniently procured, and a stone pot used, add 1 ounce of iron. The sample will melt in a good heat in about 12 minutes, if the ore is tolerably free from sulphur and iron, otherwise it will require more time.

When the sample has become quite fluid, take it out and pour it in a mould prepared to receive it, having been anointed on the inside with grease or

---

* An ounce of red lead generally contains about 1-32nd part of a grain of silver, or nearly 3 ounces of silver in a ton. Derbyshire lead ore is preferred, by some assayers, to red lead.

The proportion of silver contained in the flux must first be known, and the regular deduction made from the produce, in order to obtain a true assay.

oil ; the process of taking out and pouring the sample must be done quickly, otherwise a degree of chill will take place, so that the metal will not run freely out of the crucible, and the assay will in consequence be imperfect.

If the operation has been properly managed, the lump will separate clean from the slag or dross by a slight blow ; but if the metal and dross stick together, the assay is impure :—it is probable a little more nitre would remedy this defect.

Should the lump when broken display the metal disseminated throughout and uncombined among the slag, it is a proof the sample was not sufficiently flowed, or not kept time enough in the furnace.

If the heat is too strong, or the sample left too long time in the fire, it will set, or become dry and callous, and this change will take place to all appearance quite suddenly. Either the former circumstance of too low a heat, or this of too high, renders the assay irremediable.

Should the sample appear stubborn and refuse to melt in a brisk heat, add more nitre.

TESTING OR REFINING PROCESS.

The test or cupel should be composed of four fifths bone ashes to one fifth fern ashes, damped and well beat into an iron ring $2\frac{1}{2}$ inches deep, and 6 inches in circumference.

The test should be put in the fire an hour or more before the refining process is begun, otherwise the silver will be apt to be agitated by the unsettled test, spring over, and consequently the assay be destroyed.

Should the assay set in refining before it has become pure, throw in about half an ounce of potter's lead.*

If the fire is permitted to get low, or too much air admitted into the furnace, the assay will be apt to turn to litharge ; whenever this happens, increase the fire by putting in a few pieces of sea-coal instead of coke, at the same time sprinkle a little coal-dust on the test.

When the assay is thoroughly pure or fine it will assume a globular shape, set, or become fixed, and in a few moments will throw up sprouts or branches from the top. Take out the test, weigh the prillion, find in the table the produce or value per ton, and the work will be complete.

---

* The fire should be gradually increased toward the close of the process. A muffle or arched cover to the test would prevent the air from taking an unfavourable effect on the assay, while the furnace is opened for the purpose of increasing the fire, by adding coal, wood, or coke.

ASSAY OF COPPER ORE.

Sample — 400 grains pounded well in a mortar and sifted through a fine hair sieve, put in an earthen crucible, and frequently stirred while in the furnace with an iron rod or paddle. The sulphur will be seen to go off in white fumes; the process must be continued until this evaporation ceases, or nearly so, which will generally occupy from one to two hours. Great attention must be paid during this operation in order that a standard regal may be obtained, which being done, there will be no danger of producing a true assay. The ore, during the process, must be kept in a free, sandy state, which will be effected by stirring, and constant regulation of the degree of heat. If the ore becomes moist and begins to stick or adhere to the crucible, it must be immediately taken out of the fire and stirred a short time till this effect has ceased, and then returned. When it has become tolerably free of sulphur, it may be discovered by the evaporation having nearly ceased.* This being observed, take it out of the fire, and let it gra-

* It is only some very stubborn ores, containing a mixture of metals, or semi-metals, which require to be so effectually roasted or calcined.

dually cool in the crucible ; and if, when cold, the upper part appears red or brown, and the under part black, it is a proof of its having been well calcined.

This being done, add standard flux ; viz.

Borax . . . . . 5 dwts.
Lime . . . . . 1½ ladle.*
Fluor spar (pulverised) . . 1 ladle.

Mix these together with the calcined ore in the crucible, and cover the whole with salt,—let it melt well, and a regal will be produced.

MARKS AND REMARKS.

A good or standard regal is brown, and full of cracks or fissures, and of a spherical shape. Should it come out flat, it is a mark of its not having been well calcined, and may be thrown back again with a small quantity of nitre.

Should a regal come out too low or coarse (having, when broken, a cinder-like or cellular appearance), throw it back with additional nitre : if too high or fine (having, when broken, a metallic appearance), return it to the crucible with a ladle of sulphur ; in either case let it work well together a short time, and in all probability a standard regal will be produced.

A regal may be considered good, which will produce from 8 to 12 in 20, and this quality is

* Common assaying ladle—diameter ¾ inch, depth ½ inch.

easily known by inspection; but if less than 8, or above 12, it would be better to reject it, and begin the process again with a new sample.

Grey, black, and green ores, require a proportion of sulphur, in order to throw them back, as they contain too little of this mineral in their composition to produce a good assay.

Should a regal be too fine, put less nitre with it in refining; and therefore the coarser it is, the more nitre will be required.

FINING PROCESS.

Pound or pulverise the regal, put it in an earthen fining pot, and re-calcine it until perfectly sweet (*i. e.* free from sulphur), which may be discovered both by the appearance and fumigation. Then add

Nitre	. 3 dwts.	⎫
Red tartar	. 10 ditto	⎬ Covered or sprinkled
Borax	. 5 ditto	⎨ over with salt.
Salt	. 2 ladles	⎭

This brings down the assay into coarse copper. Should it come out having a transparent or horn-like appearance, add 4 dwts. of nitre and a ladle of salt, letting it work well in the fire. Should the assay come out black, plate it, and if the black flies off in flakes or scales, it is a proof of its not having been sufficiently calcined; if not, its colour may be attributed to lead, or a mixture of metals; the former defect renders the assay hopeless.

Should it come out clean, put the assay in the pot without flux, and when fluid, take out the pot and shake it gently until the surface assumes an azure or blue appearance ; then put

Refining flux* .   .   . 5 dwts.
(viz. 2 parts nitre, to 1 part white tartar)
Salt .   .   .   .   . 1 ladle.

Preparatory to pouring into the crucible, place the refining flux in the mouth or fore part of the scoop and the salt behind ; throw it in with the assay and let it melt until the flux settles well down, then pour the copper into one mould, and the slag or scoria into another ; return the slag into the same pot with 2 ladles of red tartar, and let it melt well down ; take out the prillion and weigh it with the lump for the produce, and the work will be completed.

---

* The refining flux should go through a calcining process before it is used; it may be done by putting 2 parts nitre to one part white tartar in an iron mortar, to which apply a red-hot iron, and stir it therewith until the deflagration has ceased ; when cold, powder and sift it.

This operation will prevent any commotion during the refining, which otherwise may be so violent as to cause some of the metal to spring out of the crucible, and thereby the assay be spoiled.

ASSAY OF LEAD ORE.

— ✦ —

Sample — 1 oz. avoirdupois.

FLUX.

1	common ladle	red tartar.
1	ditto	spar.
2	ditto	salt.
$\frac{1}{2}$	ditto	borax.
$\frac{3}{4}$	ditto	nitre.
$\frac{1}{2}$	ditto	lime.

Mix the flux with the sample and put it in an iron crucible, stir it with an iron rod during the latter part of the process; in about five minutes, in a brisk heat, the sample will be down, provided the crucible was red-hot when the assay was thrown in, which should always be the case.

If the sample, to be tried, weighs four ounces, the proportionate quantity of flux must be added, agreeably to the above statement.

It may be discovered when the sample is ready, by the grating of the rod against the bottom of the crucible in stirring—it should then be immediately taken out and poured. The metal will separate clean from the slag in a good assay.

To assay lead ore for discovering the quantity or proportion of silver it contains, the foregoing method must first be used, and the assay then tested precisely the same way as described for refining a silver sample, page 101. The lead will go off in vapour, and the silver remain in the test.

ASSAY OF TIN ORE.

———•———

Sample — Two ounces black tin.

FLUX.

Culm . . . ⅓ weight of sample.
Borax . . . 4 dwts.

PROCESS.

If the ore contains a large proportion of iron, add more culm;* when the sample is properly down, or flowed, the surface of the assay in the crucible will be perfectly smooth and motionless; in a strong heat this will occur in about twelve minutes.

When taken out of the fire, stir it well with an iron rod before you pour it; afterwards scrape the crucible, pulverise the scrapings in a mortar, and then van or wash them on a shovel. The prillion of a standard sample will not exceed 2 in 20.

The criterion for the lump is its possessing a malleable quality, or bending to the hammer without breaking.

Grain tin may be treated in every respect as the above, except in the subsequent addition of culm, which will not be required.

---

* If the sample is very stubborn, add a small quantity of pulverised flour with the culm.

METHOD OF DISCOVERING THE PROPORTION OF SILVER CONTAINED IN COPPER ORE.

Sample — One ounce.

FLUX.

1	ladle	red tartar.
1	ditto	nitre.
$\frac{1}{2}$	ditto	lime.
$\frac{1}{2}$	ditto	borax.
1	ditto	flour.
1	ditto	red lead.

Well mixed with the ore and melted in a wrought-iron crucible,* about eight minutes, in a brisk heat, will be sufficient; the last five minutes the assay should be incessantly stirred with an iron rod; pour the sample and cool it, then break out the lump, and test it in the usual way.

REMARKS.

Soon as the assay begins to flow, the lead, by the power of affinity, will presently attract the silver, or the silver, by the same law, will attach itself to the lead, and this being effected, it only requires the process of refining, or burning off the inferior metals, to find the produce.

---

* If a stone crucible be used, one ounce of iron must be added to the flux.

A TABLE

SHOWING THE NUMBER OF OUNCES AND PARTS OF AN OUNCE OF SILVER CONTAINED IN A TON OF ORE, BY ASSAY PRODUCED FROM ONE OUNCE AVOIRDUPOIS.

Assay	Per Ton			Assay	Per Ton			Assay	Per Ton		
grs.	oz.	dwts.	grs.	grs.	oz.	dwts.	grs.	grs.	oz.	dwts.	grs.
	2	6	16	4	298	13	8	8⅛	606	13	8
	4	13	8		308	0	0		616	0	0
	9	6	16		317	6	16		625	6	16
	18	13	8		326	13	8		634	13	8
	28	0	0		336	0	0		644	0	0
	37	6	16		345	6	16		653	6	16
	46	13	8		354	13	8		662	13	8
	56	0	0		364	0	0	9	672	0	0
	65	6	16	5	373	6	16		681	6	16
1	74	13	8		382	13	8		690	13	8
	84	0	0		392	0	0		700	0	0
	93	6	16		401	6	16		709	6	16
	102	13	8		410	13	8		718	13	8
	112	0	0		420	0	0		728	0	0
	121	6	16		429	6	16		737	6	16
	130	13	8		438	13	8	10	746	13	8
	140	0	0	6	448	0	0		756	0	0
2	149	6	16		457	6	16		765	6	16
	158	13	8		466	13	8		774	13	8
	168	0	0		476	0	0		784	0	0
	177	6	16		485	6	16		793	6	16
	186	13	8		494	13	8		802	13	8
	196	0	0		504	0	0		812	0	0
	205	6	16		513	6	16	11	821	6	16
	214	13	8	7	522	13	8		830	13	8
3	224	0	0		532	0	0		840	0	0
	233	6	16		541	6	16		849	6	16
	242	13	8		550	13	8		858	13	8
	252	0	0		560	0	0		868	0	0
	261	6	16		569	6	16		877	6	16
	270	13	8		578	13	8		886	13	8
	280	0	0		588	0	0	12	896	0	0
	289	6	16	8	597	6	16		905	6	16

Assay	Per Ton			Assay	Per Ton			Assay	Per Ton		
grs.	oz.	dwts.	grs.	grs.	oz.	dwts.	grs.	grs.	oz.	dwts.	grs.
12½	914	13	8	18⅜	1390	13	8	25	1866	13	8
	924	0	0		1400	0	0		1876	0	0
	933	6	16		1409	6	16		1885	6	16
	942	13	8	19	1418	13	8		1894	13	8
	952	0	0		1428	0	0		1904	0	0
	961	6	16		1437	6	16		1913	6	16
13	970	13	8		1446	13	8		1922	13	8
	980	0	0		1456	0	0		1932	0	0
	989	6	16		1465	6	16	26	1941	6	16
	998	13	8		1474	13	8		1950	13	8
	1008	0	0		1484	0	0		1960	0	0
	1017	6	16	20	1493	6	16		1969	6	16
	1026	13	8		1502	13	8		1978	13	8
	1036	0	0		1512	0	0		1988	0	0
14	1045	6	16		1521	6	16		1997	6	16
	1054	13	8		1530	13	8		2006	13	8
	1064	0	0		1540	0	0	27	2016	0	0
	1073	6	16		1549	6	16		2025	6	16
	1082	13	8		1558	13	8		2034	13	8
	1092	0	0	21	1568	0	0		2044	0	0
	1101	6	16		1577	6	16		2053	6	16
	1110	13	8		1586	13	8		2062	13	8
15	1120	0	0		1596	0	0		2072	0	0
	1129	6	16		1605	6	16		2081	6	16
	1138	13	8		1614	13	8	28	2090	13	8
	1148	0	0		1624	0	0		2100	0	0
	1157	6	16		1633	6	16		2109	6	16
	1166	13	8	22	1642	13	8		2118	13	8
	1176	0	0		1652	0	0		2128	0	0
	1185	6	16		1661	6	16		2137	6	16
16	1194	13	8		1670	13	8		2146	13	8
	1204	0	0		1680	0	0		2156	0	0
	1213	6	16		1689	6	16	29	2165	6	16
	1222	13	8		1698	13	8		2174	13	8
	1232	0	0		1708	0	0		2184	0	0
	1241	6	16	23	1717	6	16		2193	6	16
	1250	13	8		1726	13	8		2202	13	8
	1260	0	0		1736	0	0		2212	0	0
17	1269	6	16		1745	6	16		2221	6	16
	1278	13	8		1754	13	8		2230	13	8
	1288	0	0		1764	0	0	30	2240	0	0
	1297	6	16		1773	6	16		2249	6	16
	1306	13	8		1782	13	8		2258	13	8
	1316	0	0	24	1792	0	0		2268	0	0
	1325	6	16		1801	6	16		2277	6	16
	1334	13	8		1810	13	8		2286	13	8
18	1344	0	0		1820	0	0		2296	0	0
	1353	6	16		1829	6	16		2305	6	16
	1362	13	8		1838	13	8	31	2314	13	8
	1372	0	0		1848	0	0		2324	0	0
	1381	6	16		1857	6	16		2333	6	16

Assay grs.	Per Ton oz.	dwts.	grs.	Assay grs.	Per Ton oz.	dwts.	grs.	Assay grs.	Per Ton oz.	dwts.	grs.
31¾	2342	13	8	37¾	2818	13	8	44¼	3294	13	8
	2352	0	0		2828	0	0		3304	0	0
	2361	6	16	38	2837	6	16		3313	6	16
	2370	13	8		2846	13	8		3322	13	8
	2380	0	0		2856	0	0		3332	0	0
32	2389	6	16		2865	6	16		3341	6	16
	2398	13	8		2874	13	8		3350	13	8
	2408	0	0		2884	0	0	45	3360	0	0
	2417	6	16		2893	6	16		3369	6	16
	2426	13	8		2902	13	8		3378	13	8
	2436	0	0	39	2912	0	0		3388	0	0
	2445	6	16		2921	6	16		3397	6	16
	2454	13	8		2930	13	8		3406	13	8
33	2464	0	0		2940	0	0		3416	0	0
	2473	6	16		2949	6	16		3425	6	16
	2482	13	8		2958	13	8	46	3434	13	8
	2492	0	0		2968	0	0		3444	0	0
	2501	6	16		2977	6	16		3453	6	16
	2510	13	8		2986	13	8		3462	13	8
	2520	0	0	40	2996	0	0		3472	0	0
	2529	6	16		3005	6	16		3481	6	16
34	2538	13	8		3014	13	8		3490	13	8
	2548	0	0		3024	0	0		3500	0	0
	2557	6	16		3033	6	16	47	3509	6	16
	2566	13	8		3042	13	8		3518	13	8
	2576	0	0		3052	0	0		3528	0	0
	2585	6	16	41	3061	6	16		3537	6	16
	2594	13	8		3070	13	8		3546	13	8
	2604	0	0		3080	0	0		3556	0	0
35	2613	6	16		3089	6	16		3565	6	16
	2622	13	8		3098	13	8		3574	13	8
	2632	0	0		3108	0	0	48	3584	0	0
	2641	6	16		3117	6	16		3593	6	16
	2650	13	8		3126	13	8		3602	13	8
	2660	0	0	42	3136	0	0		3612	0	0
	2669	6	16		3145	6	16		3621	6	16
	2678	13	8		3154	13	8		3630	13	8
36	2688	0	0		3164	0	0		3640	0	0
	2697	6	16		3173	6	16		3649	6	16
	2706	13	8		3182	13	8	49	3658	13	8
	2716	0	0		3192	0	0		3668	0	0
	2725	6	16		3201	6	16		3677	6	16
	2734	13	8	43	3210	13	8		3686	13	8
	2744	0	0		3220	0	0		3696	0	0
	2753	6	16		3229	6	16		3705	6	16
37	2762	13	8		3238	13	8		3714	13	8
	2772	0	0		3248	0	0		3724	0	0
	2781	6	16		3257	6	16	50	3733	6	16
	2790	13	8		3266	13	8		3742	13	8
	2800	0	0		3276	0	0		3752	0	0
	2809	6	16	44	3285	6	16		3761	6	16

1

Assay grs.	Per Ton oz.	dwts.	grs.	Assay grs.	Per Ton oz.	dwts.	grs.	Assay grs.	Per Ton oz.	dwts.	grs.
50¼	3770	13	8	53¾	4013	6	16	57	4256	0	0
	3780	0	0		4022	13	8		4265	6	16
	3789	6	16	54	4032	0	0		4274	13	8
	3798	13	8		4041	6	16		4284	0	0
51	3808	0	0		4050	13	8		4293	6	16
	3817	6	16		4060	0	0		4302	13	8
	3826	13	8		4069	6	16		4312	0	0
	3836	0	0		4078	13	8		4321	6	16
	3845	6	16		4088	0	0		4330	13	8
	3854	13	8		4097	6	16	58	4340	0	0
	3864	0	0	55	4106	13	8		4349	6	16
	3873	6	16		4116	0	0		4358	13	8
52	3882	13	8		4125	6	16		4368	0	0
	3892	0	0		4134	13	8		4377	6	16
	3901	6	16		4144	0	0		4386	13	8
	3910	13	8		4153	6	16		4397	0	0
	3920	0	0		4162	13	8	59	4406	6	16
	3929	6	16		4172	0	0		4415	13	8
	3938	13	8	56	4181	6	16		4424	0	0
	3948	0	0		4190	13	8		4433	6	16
53	3957	6	16		4200	0	0		4442	13	8
	3966	13	8		4209	6	16		4452	0	0
	3976	0	0		4218	13	8		4461	6	16
	3985	6	16		4228	0	0		4470	13	8
	3994	13	8		4237	6	16	60	4480	0	0
	4004	0	0		4246	13	8				

METHOD OF COMPUTING THE VALUE OF LEAD AND SILVER ORE.

EXAMPLE.

Required the value of 16 tons 10 cwt. 2 qrs. of lead and silver ore, the produce for lead being 8⅝ in 20, and silver 3⅞ grs. from a four-ounce sample. The price of lead 22*l.* per ton, and silver 5*s.* 3*d.* per oz. Returning charges 6*l.* 10*s.* per ton, and lord's dues one twelfth for lead, and one eighth for silver.

OPERATION.

LEAD.

	tons	cwt.	qrs.
½	16	10	2
			8
	132	4	0
¼	8	5	1
	2	1	1
*20)	142	10	2

	£	s.			£	s.	d.
7	2	2	at 15 10† per ton	110	2	6
Deduct dues one twelfth				9	3	6
				100	19	0

OPERATION.

SILVER.

oz. dwt. grs.
By table (page 111) produce 3⅞ is 72 6 16 per ton (3⅞÷4=$\frac{31}{8\frac{1}{2}}$),

tons cwt. qrs.	oz. dwt. grs.	oz.	s.	d.	£	s.	d.
and 7 · 2 2 ×72	6 16=514	at 5 3			134	18	6
Deduct dues one eighth	16 17 4				118	1	2
Answer					219	0	2

* It is not usual to make any allowance for waste on rich ores.

† Returning charge deducted.

I 2

METHOD OF COMPUTING THE VALUE OF COPPER ORE.

EXAMPLE.

What is the value of 74 tons 13 cwt. 2 qrs. of copper ore, the produce, by assay, being $7\frac{1}{8}$, and standard 127l. 12s. 6d. ?

OPERATION.

$$
\begin{array}{c|rrr}
 & \pounds & s. & d. \\
\frac{1}{8} & 127 & 12 & 6 \\
 & & & 7 \\
\hline
 & 893 & 7 & 6 \\
 & 15 & 19 & 0\frac{3}{4} \\
\hline
 & 9\cdot09 & 6 & 6\frac{3}{4} \\
 & 20 & & \\
\hline
 & 1\cdot86 & & \\
 & 12 & & \\
\hline
 & 10\cdot38 & &
\end{array}
$$

THEN

$$
\begin{array}{lrrr}
 & \pounds & s. & d. \\
\text{From} & 9 & 1 & 10 \\
\text{Deduct} & 2 & 10 & 0 \quad \text{per ton returning charge.} \\
\hline
 & 6 & 11 & 10
\end{array}
$$

Again If 1 : $\overset{\text{ton}}{}$ 6 $\overset{\pounds}{11}$ $\overset{s.}{10}$ :: 74 13 2

. BY PRACTICE.

cwt.		£	s.	d.
$10\frac{1}{2}$	$\frac{1}{2}$	6	11	10
				9
		59	6	6
				8
		474	12	0
		13	3	8
		487	15	8
		3	5	11
3	$\frac{1}{7}$	0	18	10

Answer 492   0   5

Copper ores are always computed at 21 cwt. to the ton, the surplus being allowed for waste in carriage, &c.

RULE FOR DISCOVERING THE POWER OF STEAM ENGINES.

1.—Square the diameter of the cylinder, multiply the sum by ·7854* and the product by 10 :† lastly, multiply again by 144,‡ and the last product

* The established ratio of the diameter. Or look in the table, page 123, where the square inches contained in a cylinder are given, and take out the number standing opposite the given diameter.

† That is, considering the power equal to 15 lbs. to an inch, and allowing 5 lbs. or one-third for friction.

‡ Considering the stroke to be 8 feet, and the engine to go 9 strokes per minute.

will show the number of pounds the engine lifts a foot high in a minute.

2.—A horse is estimated to raise 500 lbs. 64 feet .high, or 1000 lbs. 32 feet high, or 32,000 lbs. 1 foot high in a minute, consequently, if the last product be divided by 32,000, the quotient will show the number of horses required to equal the power of the engine.

EXAMPLE.

What is the power, and horse-power of a steam engine, the cylinder being 46 inches in diameter?

$$46 \times 46 = 2116$$
$$\cdot7854$$
$$\overline{1661\cdot9064}$$
$$10$$
$$\overline{16619\cdot0640}$$
$$144$$
$$32000)\,2393145\cdot2160\,(75$$
$$224000$$
$$\overline{153145}$$
$$160000$$

ANSWER.

Engine lifts 2393145 lbs. one foot high in a minute. Equal to 75 horses, nearly.

RULE FOR DISCOVERING THE POWER OF A WATER ENGINE.

1.—Multiply the length, breadth, and depth of the bucket together, and divide by 282 (the number of cubic inches in a gallon, beer measure), multiply the quotient by $10\frac{1}{5}$ lbs. or 10 lbs. 3·2 oz. or 10·2 lbs. the weight of a gallon of water.*

2.—Multiply the diameter of the wheel by 3·1416 (the ratio of the circle), and divide the product by the circular space occupied by each bucket—the quotient will show the number of buckets contained in the wheel.

3.—Multiply the third part† of the number of buckets by the weight of water contained in one, then

4.—For the leverage—From the radius, or half the diameter, deduct the length of the crank, and one third of the remainder will be the operative length of the lever; multiply the weight of water in one third of the wheel by this length (taking the feet for the whole number of the multiplier), and the product will show the full, or entire power. Lastly, from this product cast off $\frac{1}{5}$ for friction,‡

---

* The result will be the same if the operation is done by the wine gallon, computing it to contain 231 cubic inches, and to weigh 8 lbs. 5·68 oz. or 8·355 lbs.

† There are differences of opinion respecting this; some persons contending that two-fifths the number of buckets are full at a time, but one-third is the most general and the most reasonable proportion.

‡ Here again engineers are not unanimously agreed, some

and the remainder will show the net or real power of the wheel.

EXAMPLE.

Required the power of a water-wheel, the diameter being 46 feet, the buckets 30 inches long, 12 inches deep, and 6 inches wide, with $1\frac{1}{4}$ inches between each bucket, and the crank 3 feet long.

OPERATION.

1.—To find the quantity and weight of water in each bucket.

$$30 \times 12 \times 6 = 2160 \div 282 = 7\cdot66 \times 10\cdot2 = 78\cdot13.$$

2.—To find the number of buckets contained in the wheel.

$$46 \times 12 = 552 \times 3\cdot1416 = 1734 \ (6 + 1\tfrac{1}{4}) \div 7\cdot25 = 239.$$

3.—To find the weight of water on $\frac{1}{3}$ of the wheel.

$$239 \div 3 = 80 \times 78\cdot13 = 6250.$$

4.—To find the power of the lever.

$$46 \div 2 = 23 - 3 = 20 \div 3 = 6\cdot66$$

and

$$6250 \times 6\cdot66 = 41625.$$

allowing one-fifth, some one-fourth, and some even one-third of the power, for friction. It is true the distance the wheel is placed from the work, and other contingent circumstances, must be taken into consideration; but in ordinary operations, where the wheel draws close, one-fifth is very near the truth.

Lastly, for friction.
5) 41625
   8325

Answer, 33300 lbs. the actual power of the wheel.

To find the depth at which a wheel will draw a column of water in a lift of pumps of any given dimensions.

RULE.—Find the power of a wheel by the foregoing method, then from the table, page 125, take out the weight of water in a fathom of the given size pump. Divide the power of the wheel (in pounds) by this number, and the quotient will show the fathoms.

EXAMPLE.

The power of the fore-mentioned wheel is 33300 lbs., how deep will she draw in a 12-inch lift of pumps?

294·53) 33300 (113

Answer 113 fathoms.

TO FIND THE HORSE-POWER OF A WHEEL.

RULE.—Multiply the power (found by the given rule, page 119) by the number of revolutions made by the wheel in a minute, and this product by the length of the stroke in feet, or double the length of the crank: divide the last product by 32000, and the quotient will show the number of horses required to equal the power of the wheel.

The fore-mentioned wheel is allowed to make seven revolutions in a minute; required the horse-power.

$$
\begin{array}{r}
33300 \\
7 \\
\hline
233100 \\
6 \\
\hline
\end{array}
$$

$$
32{\cdot}000)\overline{1398{\cdot}600}(44 \\
\phantom{32{\cdot}000)}128 \\
\phantom{32{\cdot}000)}\overline{118} \\
\phantom{32{\cdot}000)}128 \\
\phantom{32{\cdot}000)}\overline{}
$$

Answer, 44 horse-power, nearly.

TABLE

Diameter of Cylinder	Square Inches	Diameter of Cylinder	Square Inches	Diameter of Cylinder	Square Inches	Diameter of Cylinder	Square Inches
10	78·54	26	530·93	42	1388·59	58	2642·00
11	95·03	27	572·56	43	1452·20	59	2734·00
12	113·10	28	615·75	44	1520·53	60	2827·44
13	132·73	29	660·20	45	1590·43	61	2922·47
14	153·94	30	706·86	46	1661·91	62	3019·00
15	176·71	31	754·77	47	1735·00	63	3117·25
16	201·06	32	804·25	48	1809·56	64	3217·00
17	226·98	33	855·30	49	1885·74	65	3318·31
18	254·47	34	907·92	50	1963·50	66	3421·20
19	283·54	35	962·00	51	2042·82	67	3526·66
20	314·16	36	1017·88	52	2123·72	68	3651·69
21	346·36	37	1075·20	53	2206·19	69	3739·29
22	380·13	38	1134·00	54	2290·23	70	3848·46
23	415·47	39	1194·60	55	2375·83	71	3959·20
24	452·39	40	1256·64	56	2463·00	72	4071·51
25	490·88	41	1320·26	57	2651·76	73	4185·40

NOTE.

The annexed table of the quantity and weight of water contained in 6 feet of pump (page 125) may be proved or extended by the following rules; viz. Square the diameter of the pump, and multiply the product, first by the decimal ·7854, again by the length of the pump in inches, and divide by 231; the whole numbers in the quotient will show the wine gallons. Then, to find the cubic feet, divide the solid inches by 1728. Again, to find the weight, multiply the cubic feet by 1000, and divide the product by 16.

How many pounds, wine gallons, and cubic feet, are contained in a cylinder, or pump, 12 inches in diameter, and 6 feet in length?

TO FIND THE WINE MEASURE.

$$12 \times 12 = 144 \text{ Square of diameter.}$$
$$\underline{\cdot 7854} \text{ Multiplier.}$$
$$113 \cdot 0976$$
$$\underline{\quad 6}$$
$$678 \cdot 5856$$

Cubic inches in   12

a wine gallon = 231)8143·0272(35·2512

Answer, 35 gals. 1 qt. $\underline{\quad\quad 4}$ $1\cdot0048$ ·

TO FIND THE CUBIC FEET.

Inches

in a cubic foot—1728)8143·0272 Inches as before.

$\underline{4\cdot7124}$ Cubic feet.

TO FIND THE POUNDS.

$4\cdot7124$ Cubic feet as before.

$\underline{\quad 1000}$ oz. weight of a cubic foot of water,

16)4712·4000

$\underline{294\cdot525}$ lbs. weight.

A TABLE

SHOWING THE WEIGHT, WINE GALLONS, AND CUBIC FEET OF WATER CONTAINED IN SIX FEET OF PUMP, FROM FOUR TO TWENTY INCHES IN DIAMETER.

Diameter of Pump	Weight	Wine Measure			Cubic Feet	Diameter of Pump	Weight	Wine Measure			Cubic Feet
ins.	lbs.dec.	gal.	qts.	pts.	ft. dec.	ins.	lbs.dec.	gal.	qts.	pts.	ft.dec.
4	32·75	3	3	1	·522	12¼	306·95	36	2	1	4·910
4¼	36·95	4	1	1	·591	12½	319·60	38	1	0	5·113
4½	41·42	4	3	1½	·662	12¾	332·51	39	2	1	5·319
4¾	46·15	5	2	0	·738	13	345·68	41	1	1	5·530
5	51·14	6	0	1	·818	13¼	359·10	42	3	1	5·745
5¼	56·38	6	3	0	·902	13½	372·78	44	2	1	5·960
5½	61·87	7	1	1½	·989	13¾	386·72	46	0	1	6·187
5¾	67·63	8	0	0½	1·082	14	400·90	48	0	0	6·414
6	73·63	8	3	0	1·178	14¼	415·35	49	2	1	6·645
6¼	79·90	9	2	0	1·278	14½	430·00	51	1	1	6·880
6½	86·42	10	1	0	1·382	14¾	445·00	53	0	1	7·119
6¾	93·20	11	0	0	1·491	15	460·23	55	0	1	7·363
7	100·22	12	0	0	1·603	15¼	475·69	56	3	1	7·610
7¼	107·51	12	3	0	1·720	15½	491·42	58	3	0	7·862
7½	115·00	13	3	0	1·840	15¾	507 40	60	2	1	8·117
7¾	122·85	14	2	1	1·965	16	523·63	62	2	1	8·379
8	130·90	15	2	1	2·094	16¼	540·13	64	2	1	8·641
8¼	139·22	16	2	1	2·227	16½	556·87	66	2	1	8·909
8½	147·78	17	2	1	2·354	16¾	573·88	68	2	1	9·181
8¾	156·60	18	2	1	2·505	17	591·13	70	3	0	9·457
9	165·68	19	3	0	2·650	17¼	608·65	72	3	0	9·739
9¼	175·00	20	3	1	2·800	17½	626·42	75	0	0	10·022
9½	184·60	22	0	0	2·953	17¾	644·67	77	0	1	10·310
9¾	194·45	23	1	0	3·110	18	662·73	79	1	0	10·602
10	204·54	24	1	1	3·272	18¼	681·26	81	2	0	10·899
10¼	214·90	25	2	1	3·438	18½	700·00	83	3	0	11·142
10½	225·51	27	0	0	3·607	18¾	719·10	86	0	1	11·504
10¾	236·37	28	1	0	3·781	19	738·40	88	1	1	11·813
11	247·50	29	2	1	3·959	19¼	757·96	90	3	0	12·126
11¼	258·87	30	3	1	4·141	19½	777·78	93	0	1	12·443
11½	270·51	32	1	1	4·327	19¾	797·85	95	2	0	12·764
11¾	280·40	33	2	1	4·518	20	818·18	97	3	1	13·090
12	294·53	35	1	0	4·712						

A TREATISE

ON THE QUALITY, MANUFACTURE, AND CHOICE OF
CORDAGE, FOR MINING PURPOSES, WITH RULES AND
TABLES FOR THE WEIGHT AND NUMBER OF THREADS
CONTAINED IN ANY SIZE ROPE.

It is certainly very desirable, if not absolutely necessary, that every person who is intrusted with
the management of a mine should possess some
means of obtaining, with a degree of certainty, the
quality and weight of the ropes he may have occasion to use ; otherwise the lives and property intrusted to his care will be continually placed in
jeopardy, and his employers be always subject to
impositions respecting the charge ; because in
many cases (from the magnitude of the material) it
cannot be weighed, and therefore its weight can
only be ascertained by computation ; consequently,
if the agent is ignorant of the matter, the right of
the adventurers will solely depend on the truth of
the manufacturer's calculation.

The following tables will enable the agent to find
the weight of any rope, and the ensuing remarks
will help his judgment respecting the quality
thereof ; being far the most important part of the
subject.

There are various methods of discovering the
quality of hemp ; but as miners have seldom an
opportunity of inspecting the article in this stage of

preparation, we shall pass on, and show how it may be proved after its having been completely manufactured.

The first thing that demands our particular attention is, the size of the yarn or thread of which the rope is composed. There is a certain gauge or standard for this, known among ropemakers by the terms, sixteens, eighteens, twenties, &c., which means 16, 18, or 20 yards in the strand, or third part of a rope 3 inches in circumference. The following table shows the weight of the different sizes of yarn before it has gone through the operation of tarring.

SIZE	LENGTH	WEIGHT
		lbs. oz.
25		2 13
20		3 8
18	170 fathoms.	3 15
16		4 6
15		4 10

Now the true standard size for shroud-laid rope is *twenties*,* and it is of consequence that agents should give their orders accordingly, and afterwards be assured that their ropes have really been made with yarns of this gauge.

In order to prove this, first, girt the circumference of the rope, then count the yarns in the

* Of which it is shown in the above table that 170 fathoms weighs only 3 lbs. 8 oz. or 3½ lbs.

strand, and, lastly, refer to the table (page 135), and note if the number corresponds with that standing in the proper column, opposite the dimensions of the rope.

Manufacturers have many inducements for spinning their yarn large. First, It is less expensive, for it requires no more time to spin a large yarn than to spin a small one, and 16 or 17 yarns (in their way) will answer the end of 20. Secondly, In large yarns, inferior or refuse hemp can be spun, which cannot be done in yarns of a smaller size ; and this consideration, if there was no other, should cause the agent to be exceedingly particular in having his rope made of standard yarns ; and let it be remarked, that although a rope made of sixteens or eighteens will be nearly equal in weight to another made of twenties, yet by no means will it be equal in strength, even if made of the very same kind, or indeed of superior hemp. This is too plain a truth to need any illustration : for though it may be argued that what is wanting in number is made up in bulk, yet it will support an equal weight no more, in proportion, than a body of raw hemp the size of a cable will be as strong as the cable itself.

By inspecting the table (page 135) it will be seen that the strand of a 16-inch capstan-rope made of twenties contains 569 yarns, but if made of sixteens, only 455 yarns ; making a difference in the whole rope of 342 yarns.

We shall now give a plain and expeditious, though infallible, method, of proving the quality of

hemp and yarn, viz. from the end, or *fag*, of the
rope cut several of the yarns in fathom lengths,
each of these (standard. size) should suspend,
or bear up separately, 70 pounds weight at the
least.

Regard must next be paid to the last part of the
manufacture, called the *lay*, or twist, of the rope;
and this should undergo a strict examination, as
much depends on the skill and attention of the manufacturer in this part of the process; for it is very
possible that the best materials may be used, the
yarn spun of the proper size, and with the greatest
care, and yet the rope be very defective, and by no
means fit to be depended on. This may be easily
discovered when the rope is laid in a straight line;
then, if either of the strands is observed to mount
or fall,* that is, rise above or sink beneath the
others in any degree, the rope has been *crippled*, or

* This fault or defect is known among ropemakers by the
term 'pinch;' and as the remedy occasions a great deal of
trouble and delay, it is too often suffered to pass, especially as
few persons are able to detect it or are aware of its injurious
tendency.

There are many casual occurrences whereby ropes are exposed to injuries in mines, out of the common course of working.
We may notice an instance or two, viz. inattention or ignorance
in taking them from the coil when new; they should always be
taken out the contrary way from which they were coiled in; that
is, if a capstan rope is coiled into a waggon, the uppermost end
should be put down and drawn from the under part of the carriage; also, in small cordage, the inward extremity of the rope
should be taken and drawn through the aperture of the reel.
The general disproportion of capstan gear in mines has a most
destructive effect on the ropes; the sheaves or *pulleys*, as well as
the *barrel*, of the capstan being considerably too small; indeed,
there is still room for much improvement in this part of mining
machinery.

K

inevitably spoiled ; for if the former case, of one strand rising, in the event of trial, that strand will be found to bear little or none of the weight, when the other two will break ; and in the latter case, of one strand sinking, that strand will break before the other two have been brought to the strain, or have borne any considerable part of the weight.

These great defects in cordage are too often to be found, and almost as often pass unobserved ; but they may always be detected by a close inspection, and thereby many of the serious injuries and fatal accidents which so often take place in mining be happily prevented.

We shall close these observations after remarking, that as nearly all cordage used in mining is much exposed to the alternate influence of sun and moisture, which tends greatly to accelerate its decay, it ought by all means to contain a greater quantity of tar than is generally used. The common rule is 1 to $5\frac{1}{2}$, or 1 to 6 ; but the proportion of 1 to $4\frac{1}{2}$, or 1 to 5, would be much better : but we recommend this increase for standing-ropes only, such as capstan-ropes, &c. ; as from the comparative unfrequency of their use, and the length of time they endure, they are equally liable to injury from mould and decay, as from strain and friction.

The common practice of tarring the surface of the rope after it has been manufactured is of very little service : the way we recommend is, by reducing the ordinary weight suspended to the lever, during the process of tarring the yarn in the manufactory, when it is drawn in a body from the heated coppers through the knipper, whereby the tar being lodged in the internal part of the rope cannot fail of preserving it under all circumstances. The following rules, examples, and tables will be found plain, convenient, and correct.

TO FIND THE NUMBER OF THREADS IN A SHROUD-LAID* ROPE.

RULE.—State the question as in direct proportion, square the first and third terms, multiply the second and third terms together, and divide the product by the first.

How many standard yarns, or threads, are there in a 14-inch capstan rope?

$$\begin{array}{ccc} \text{in.} & \text{yarns.} & \text{in.} \\ \text{As } 3 & : \ 20 \ :: & 14 \\ 3 & & 14 \\ \hline 9 & & 56 \\ & & 14 \\ & & \overline{196} \\ & & 20 \\ & & \overline{9)3920} \end{array}$$

Answer 435 threads in the strand.
$$\frac{3 \cdot}{}$$
or 1305 threads in the rope.

* The term 'shroud-laid' is used to distinguish a rope of three strands or parts from another of nine strands, which is termed 'cable-laid.' The latter may be said to be 3 shroud-laid ropes twisted together. It is seldom that any other but three-strand ropes are used in mines.

How many standard yarns are there in a $9\frac{1}{2}$-inch rope?

```
    in.    yarns    in.
As 3  :   20   :: 9·5
    3             9·5
   ─             ───
    9             475
                 855
                ─────
                90·25
                   20
            ─────────
         9)1805·00
    Answer   200· threads in the strand.
              3
            ───
            600 threads in the rope.
```

TO FIND THE WEIGHT OF SHROUD-LAID ROPES.

RULE.—State the question and square the numbers as in the last example.

EXAMPLE.

If 1 cwt. of 3-inch rope measures 54 fathoms,* what will be the length of an cwt. of a 12-inch rope?

* The length of standard yarn to a lb. is 43 fath. 2 ft. 4·3 in. after it has been tarred, and 54 fathoms of 3-inch rope are exactly 1 cwt. By this rule the following tables have been constructed. It must be recollected that this computation is made, estimating the proportion of tar 1 to $5\frac{1}{2}$ only.

```
       in.          fath.           in.
As      3      :     54      : :    12
        3            9               12
       ──          ─────            ───
        9      144)486(3·375        114
                   432      6
                  ─────    ────
                   540    2·250
                   432      12
                  ─────    ─────
                  1080    3·000
                  1008
                  ─────
                   720
                   720
                  ─────
```

fath. ft. in.

Answer 3   2   3

If 100 fathoms of 3-inch rope weighs 1 cwt. 3 qrs. 11·3 lbs., what will be the weight of a 15-inch rope the same length?

```
      in.   cwt.  qrs.  lbs.        in.
As     3  :  1     3    11·3  : :   15
       3     4                      15
      ──    ──                     ───
       9     7                      75
            28                      15
           ─────                   ───
           207·3                   225
            225
          ────────
        9)46642·5
      112) 5182  (46
           448
          ─────
           702
           672
          ─────
            30
          ═════
```

cwt. qr. lbs.

Answer 46   1   2

TO FIND THE WEIGHT OF ROPES BEING 120 FATHOMS IN LENGTH.

RULE.—Divide the circumference of the rope by 2, and square the remainder.*

EXAMPLE.

What is the weight of a 12-inch rope 120 fathoms long?

$$2\overline{)12}$$
$$\overline{6}$$
$$\underline{6}$$

Answer $\underline{\underline{36}}$ cwt.

EXAMPLE.

What is the weight of a 14½-inch rope 120 fathoms long?

$$2\overline{)\ 14{\cdot}5}$$
$$\overline{7{\cdot}25}$$
$$\underline{7{\cdot}25}$$
$$\overline{52{\cdot}5625}$$
$$\underline{4}$$
$$\overline{2{\cdot}2500}$$
$$\underline{28}$$
$$\overline{200}$$
$$\underline{50}$$
$$\overline{7{\cdot}0000}$$

                 cwt. qrs. lbs.

Answer   52   2   7

* This rule is not perfectly accurate, but may be useful in affording a clue for finding the approximate weight of ropes, especially if the circumference is in even numbers; it may then be used mentally, or by the mind only.

TABLE I.

SHOWING THE NUMBER OF THREADS IN THE STRAND
OF A SHROUD-LAID ROPE.

Size of Rope	Size of Yarn Sixteens	Size of Yarn Twenties	Size of Rope	Size of Yarn Sixteens	Size of Yarn Twenties
Inches	Number	Number	Inches	Number	Number
2	7	9	9½	160	200
2½	10	14	10	177	222
3	16	20	10½	196	245
3½	21	27	11	215	268
4	28	35	11½	235	293
4½	36	45	12	256	320
5	44	55	12½	278	347
5½	53	67	13	300	375
6	64	80	13½	324	405
6½	75	93	14	348	435
7	87	109	14½	374	467
7½	100	125	15	400	500
8	113	142	15½	427	533
8½	128	160	16	455	569
9	144	180	16½	484	605

The difference of the number of threads in the whole rope will be found by subtracting the numbers standing under 'sixteens' from those under 'twenties,' and multiplying the remainder by 3. (Thus 256 — 320 = 64 × 3 = 192.) Therefore it appears that a 12-inch rope made of 'twenties' contains 192 threads more than another, of the same circumference, made of 'sixteens.'

TABLE II.

SHOWING THE LENGTH OF SHROUD-LAID ROPES TO A
CWT. FROM 1 TO 16 INCHES IN CIRCUMFERENCE.

Size of Rope	Fath.	Feet	Inches	Size of Rope	Fath.	Feet	Inches
Inches				Inches			
1	486	0	0·0	9	6	0	0
1½	216	0	0·0	9½	5	2	3·7
2	121	3	0·0	10	4	5	1·9
2½	77	4	6·7	10½	4	2	5·4
3	54	0	0·0	11	4	0	1·2
3½	39	4	0·0	11½	3	4	0·9
4	30	2	3·0	12	3	2	3
4½	24	0	0·0	12½	3	0	7·9
5	19	2	7·7	13	2	5	3·1
5½	16	0	4·3	13½	2	3	11·5
6	13	3	0·0	14	2	2	10·4
6½	11	3	0·0	14½	2	1	10·4
7	9	5	6·2	15	2	0	11·5
7½	8	3	10·0	15½	2	0	1·4
8	7	3	6·7	16	1	5	4·6
8½	6	4	4·3	16½	1	4	8·5

The exact weight of a rope of any length may be found by the
above table and the rule of proportion.

EXAMPLE.

Required the weight of 25 fathoms of 14-inch rope?

```
        fath. ft.  in.        cwt.    fath.
    As    2   2   10·4    :    1   ::   25
          6                              6
         ──                            ───
         14                            150
         12                             12
        ────                          ────
        178·4                   178·4)1800(10·09
                                          4
                                        ────
         cwt. qrs. lbs.                  36
Answer    10   0   10                    28
                                        ────
                                        10·08
```

TABLE III.

SHOWING THE WEIGHT OF SHROUD-LAID ROPES 100 FATHOMS IN LENGTH.

Size of Rope	Cwt.	Qrs.	Lbs.
Inches			
3	1	3	11·3
4	3	1	4
5	5	0	15
6	7	1	16
7	10	0	7
8	13	0	19
9	16	2	15
10	20	2	4
11	24	3	11
12	29	2	14
13	34	3	3
14	40	1	12
15	46	1	2
16	52	2	22

NOTE.—The weight of any length of rope may be found by the above table and the rule of practice.

EXAMPLE.

What is the weight of 47 fathoms of 11-inch rope ?

cwt. qrs. lbs.

25	$\frac{1}{4}$	24	3	11
		6	0	24
20	$\frac{1}{5}$	4	3	24
2	$\frac{1}{10}$	0	1	27
Answer		11	2	19

The weight and circumference of any rope being given, the length may be found by the foregoing table and the common rule of proportion, or by decimals.

EXAMPLE.

cwt. qrs. lbs.

What is the length of 7   3   14 of 7-inch rope ?

```
    cwt. qrs. lbs.      fath.        cwt. qrs. lbs.
As  10   0   7    :    100    ::    7   3   14
     4                                 4
    ──                                ──
    40                                31
     4                                 4
   ───                               ───
   161                               126
                                     100
                                  ──────────
                             161)12600
                                  ──────────
                                    78·26
```

OTHERWISE

```
   cwt.                 fath.         cwt.
As 10·0625      :       100    ::    7·875
                                      100
                                  ──────────
                          10·0625)787·500
                                  ──────────
                                    78·26
                                        6
                                     ──────
                                     1·56
                   fath. ft. in.       12
                                     ──────
           Answer   78   1   6        6·72
```

Or the length of any rope may be known by the 2nd table and the rule of practice.

EXAMPLE.

What is the length of 16 cwt. 2 qrs. 21 lbs. of a 10½-inch rope?

```
            fath. ft.  in.
      ½  |   4    2   5·4  to an cwt. by the table.
         |            8
         |  ──────────────
         |  35    1   7·2
         |            2
         |  ──────────────
         |  70    3   2·4
      ¼  |   2    1   2·7
      ½  |   0    3   3·6
         |   0    1   7·8
         |  ──────────────
 Answer     73    3   4·5
```

OBSERVATIONS

ON THE CONSTRUCTION OF MINING CAPSTAN AND SHEARS.

IT has been noted that there remains much room for improvement in the capstan machinery of our mines; consequently it will not be irrelevant to the foregoing treatise briefly to submit our ideas on the subject in this place.

By the present apparatus, a capstan rope of 14 or 16 inches in circumference is drawn over a single pulley, on the top of the shears, two feet in diameter, or thereabout, from whence it is brought down nearly to the foot, and then conducted under another pulley to the axle or barrel of the capstan; a cylinder seldom exceeding $2\frac{1}{2}$ feet in diameter.

The violence done to the rope in lifting great weights, as well as the augmentation of friction, in consequence of these sudden turns or incurvities, is certainly greater than has been generally understood; but reflection and experience combine to teach us, that at all these bends or flexures a considerable proportion of the number of threads contained in the rope contribute nothing towards the support of the suspended weight; the outermost part of the cordage, from the grove of the pulley, necessarily bearing the whole strain.

The excess of friction is chiefly produced by the

lay or twist of the rope, it having thereby an in-
nate and unconquerable tendency to resist the
constraint of being forced into a small curve; consequently, the less scope it has the more power will
be required.

These arguments are self-evident, and no doubt
but the inconveniences alluded to, and many more,
are partially known and acknowledged; but the
question is, How can they be remedied? We respectfully submit the following proposition :—First, let
the uprights of the shears stand farther apart on
the top than usual, and the cap or head-piece of a
convex shape, so that it may admit 3 pulleys in a
triangular position,* the side pulleys horizontal,
and the middle one standing higher than the others;
then let a sheave be introduced in a channel in the
leg at the shears next the capstan, about one third
down from the cap; another in the stay of the span-
beam, projecting 3 or 4 feet from the leg; and the
last pulley placed in a strong post fixed about 14
feet from the foot of the shears, in a direct line and
· parallel with the axle of the capstan.†

By this simple contrivance it may be reasonably
conjectured that one rope will endure as long as
two, and a 14 be equal to a 16-inch on the old plan;
and as the latter, 100 fathoms long, is above 12 cwt.
heavier than the former, of course that superfluity

* Or a straight piece may be used for the cap, letting the
centre pulley be large and the side ones small; but a semicir-
cular cap with 5 small pulleys (one on the top and two on each
side) will be still preferable.

† The barrel of the capstan should be made proportionally
larger, and the arms or bars of a corresponding length.

will be avoided both in cost and gravity: * the shears will stand more erect than usual, whereby their height and strength will evidently be increased.

We profess to know but little of the science of mechanics, nevertheless we feel confident that the foregoing proposition is quite practicable, and the subject is sufficiently important to demand consideration. We hope to see the hint improved and carried into effect by some of our ingenious engineers.

* The safety, as well as convenience and economy, of this plan is too apparent to be overlooked.

SUPPLEMENT,

OR

THIRD PART

OF THE

PRACTICAL MINER'S GUIDE.

PRACTICAL MINER'S GUIDE.

PART III.

INTRODUCTION.

NEARLY twenty years have elapsed since 'The Practical Miner's Guide' was sent into the world, and the author is constrained to acknowledge that his extensive practice and experience in almost every branch of mining during that long period has not enabled him to discover where any important improvement can be made in that work. The *additions* in the present volume are comprised in merely extending the principles originally laid down, and practically applying the mathematical tables and rules to more difficult, complicated, and momentous subterranean surveys; for it may be remarked, without subjecting ourselves to the charge of arrogancy, that the *light* thrown on the mining world by this publication has contributed towards expanding the minds of practical men generally, and qualifying them for pursuing and carrying out the scientific and demonstrative principles to the utmost extent of their difficult and highly important operations.

During the last five years the author commenced a correspondence on mine surveying in the 'Mining Journal,' with a view of rendering a lasting benefit to the mining interest, by exposing the imperfection of the old method of dialling, by *tracing*, *pegging*, or mechanical repetition, and showing the vast superiority of the trigonometrical system in every point of view. This attempt brought on a most extensive controversy, and many of 'the old school' arose and came forward to advocate the old practice; but, as if ashamed of the cause they had espoused, they were, almost to a man, anonymous writers.

The unprejudiced and investigating part of the community honoured me, or rather the system I defended, with their able support, and rallied around my standard. The defeat of our opponents was complete, even before we attempted to apply the infallible test of *experimental operation*. Reason and historical proof had thrown them down; but when we brought forward problems, in the shape of actual surveys, and challenging them to furnish the true answers, they crept out of the field, and not one of them was heard of after that interesting, beneficial, and satisfactory investigation was introduced. Nearly all the problems that appear in this supplement were publicly brought forward by the author on this occasion, and of course underwent a rigid and extensive scrutiny; and, so far from any error appearing, the truth of the calculation and perfection of the system was demonstrated by exact corresponding solutions being publicly recorded from persons residing in England, Ireland, and Wales.

It will be seen, in the preface to the first edition of this work, that I called public attention to the great injury done to mining by the bad practice of surveying or dialling then in general use; but as at that time there was no work written on the subject, there was an excuse for the disastrous errors that took place then, that does not now exist; and the mine agent who cannot now *prove* all his dialling operations, before a single stroke has been struck, ought not to presume to make the attempt; and certainly it behoves directors, managers, and shareholders in mines, to test the abilities of those agents or captains who profess to be competent to undertake the momentous work of subterraneous surveying, by requiring them to give mathematical solutions to a series of practical questions on the subject; and not, as is too often the case, proving their incapacity by some fearful error that they have committed, occasioning serious injury to the mine, great delay, and an extra cost of hundreds of pounds to the proprietors. And let it be understood that *proof* cannot be obtained by the old method of dialling; that is, by tracing or repeating at the surface the drafts taken underground. This was fully demonstrated in the course of the late controversy, both argumentatively and practically. One convincing case was brought forward and confirmed, that occurred in Gwennap several years ago. It was required to find the point at surface for sinking a shaft to meet a rise that had been

commenced at the adit level. This was certainly
a very plain and easy job under a good system, but
it presented insurmountable difficulties under the
old practice. The shaft was set to a pair of eight
men, and the captains proceeded to *dial* and *peg* to
ascertain the true point. The first dialler fixed
his terminating peg on a certain spot, and the
next dialler's peg occupied another position; and
the history of the case is, that four agents were
constantly occupied two or three weeks about this
job, without being able to ascertain the true place
for the shaft, which a competent man would have
determined and proved in a few hours. At the
close, they had a plantation of pegs occupying an
area of several square fathoms; and the manager,
seeing it hopeless to expect *certainty* from such an
uncertain practice, and tired of waiting and wasting
any more time, proceeded to make the best of a bad
matter, by lining out the shaft in the place where
the pegs appeared to stand thickest! and the consequence was, a serious error of some feet in the
holing.

Now, by introducing this affair, our design is
not to impeach *men*: it is the *practice* or usage
that we condemn and expose, in order that it
may be discountenanced and rejected, and thereby
an incalculable benefit be conferred on metallic
mining.

TRAVERSE DIALLING.

IT is well known that our little army of mine agents or captains have almost to a man been selected from the 'ranks,' and have been brought up as working miners from their youth ; and the best judges admit the propriety and even necessity of this regulation, in order that they should be competent to fix the value or fair price for working a tut-work bargain, or tribute pitch, and possess a thorough knowledge of underground operations. This being admitted, we must conclude, that in general their education has been much limited ; and, therefore, in writing for their assistance, although there may be some exceptions, we would accommodate ourselves to the lowest capacity, so that the young aspiring miner may not be discouraged from prosecuting his incumbent and laudable studies, and qualifying himself for performing the high and paramount duties of a mine agent with credit to himself, and advantage to his employers and his country.

Before this work was published, it is questionable whether a mining traverse had ever been trigonometrically solved in this country ; and, consequently, in order to simplify the matter, the instructions given for finding the ultimatum of a course of dialling was by construction or instrumental operation; but as we believe our students are generally prepared to advance a step, we shall now recommend and show the more excellent way of performing the whole by computation, or by figures.

The trigonometrical method of working a course of dialling reduces the whole, however numerous and diversified, into two numbers; for the four columns of easting, westing, northing, and southing being added up separately, and then the less deducted from the greater of the opposite cardinal points, reduces the whole into two numbers, forming the base and perpendicular of the great triangle, and are necessarily right angle cardinal bearings, such as easting and southing, or northing and westing, as the case may be; and our next and last operation is to find the hypothenuse and angle corresponding with those two sides, which hypothenuse and angle is the final line, or course of the survey.

·   EXAMPLE.

A traverse has been worked, the columns added up, and the westing subtracted from the easting, showing the excess of easting to be 346 feet; and the southing subtracted from the northing, the difference proved the excess of northing, 419 feet 5 inches.

Find the hypothenuse by square root.

RULE.—Add the sum of the squares of the two sides together, and extract the square root of their sum.

$$346 \times 346$$

```
    346
    346
  ─────
   2076
   1384
  1038
 ──────
 119716
```

```
    419·4
    419·4
  ───────
    16776
    37746
     4194
    16776
 ──────────
 175896·36
 119716
 ──────────
 295612·36(543·7
 25                12
 ──────          ────
 104) 456          8·4
      416
 ──────────
 1083) 4012
       3249
 ──────────
 10867) 76336
        76069
 ──────────
        267
```

ft.   in.

Answer Hyp. 543   8

Find the angle by proportion.

ft.　in.　　　　ft.　　　　　　　　ft.

If 419　5　gives　346　what will　6　give?

　12　　　　　　12　　　　　　　12

　5033　　　　4152　　　　　　72

　　　　　　　　72

　　　　　　8304

　　　　　29064　　in.　　ft.　in.

　5033)298944(59·39　or　4　11·39

　　　25165

　　　47294

　　　45297

　　　19970

　　　15099

　　　48710

　　　45297

　　　4413

Then by inspection in the second table, page 79, this quotient of 4 feet 11·39 inches will be found standing opposite 39° 30′, which is the bearing, or sum of the angle opposite the shortest side of the great triangle.

ANSWER.

Hypothenuse, or direct length from beginning to end, 543 feet 8 inches. Bearing, or direction from beginning to end, 39° 30′ east of north.

REMARKS.

In carrying out this system practically, after we have laid down this grand or final line at surface, and fixed a mark at the extreme end of the line which has been measured off from the starting point, 543 feet 8 inches on the bearing, 39° 30' east of north (or 50° 30' north of east, the complement), we are furnished with a double means of proving if this length and angle has been correctly laid down, by measuring off, due north, 419 feet 5 inches from the start, and then placing the theodolite, or dial, on the end of that line, and measuring off due east 346 feet; consequently, if the whole has been well done, the last mark will exactly agree in both cases. Or should the ground be more favourable, we may avail ourselves of the convenience of laying off the east line first, and the north line last, which will bring us to the same point.

One great advantage of these proof lines will appear, when we take into consideration that most of the instruments used in mines for taking horizontal angles have no vernier scale for reading off the fraction of the angle ; and, therefore, if the bearing falls between any quarter, or half of a degree, the surveyor must depend on the judgment of his eye for the division, and let it be known that an error of one quarter of a degree in 100 feet amounts to 5 inches and a decimal of ·23596, or upwards of 2 feet 7 inches in a line of 100 fathoms ; hence the value of having this most satisfactory and convenient check for the laying down of the last grand line must be manifest to every observer, and should never be neglected.

LOGARITHMS.

Should the practitioner wish to prove the finding the angle and hypothenuse by logarithms, the following is the rule :—

From less side 346 and radius   = 12·5390761
Subtract longest side, 419·4   = 2·6227140
Log. tangent of 39° 30'   .   . 9·9163621

RULE FOR THE HYPOTHENUSE.

From less side and radius (as before)   12·5390761
Subtract sine of 39° 30'   .   .   .   9·8035105
Logarithm of 543·8 nearly   .   .   2·7355656

The rules expressed at length read thus :—

FOR THE ANGLE.

Add the radius to the logarithm of the less side, and from the sum subtract the logarithm of the greatest side ; the remainder, or sum, will be the tangent of the angle opposed to the less side.

FOR THE HYPOTHENUSE.

Add the logarithm of the given side to the sine of the angle opposite to the side required, and from the sum subtract the sine of the angle opposed to the given side ; the remainder will be the logarithm of the side required.

SYSTEM.

There is much propriety in the remark, that ' system is the handmaid of science,' and the term may be considered as used in contradistinction to disorder, irregularity, or random. The man who would excel in the important work of mine surveying should have a *system*, and a good one. It is true, men are apt to be bigoted in this matter, and think so highly of their own system as to despise all others ; but certainly we must admit that a bad or an imperfect system is better than no system at all. He who has no fixed rule is liable to error every step he takes. We would recommend the young dialler to adopt a system in keeping his register or dialling book underground, so that his subterranean surveys may be perfectly clear and comprehensible, not only to himself, but to all practical men. Let us suppose we have to survey a level driven on the course of the lode, where there are several cross cuts driven off to the right and left. I would advise the student to keep the number of his drafts on the main line, or course of the lode, in regular numerical order ; and when he has to branch off on a cross cut, let him make the necessary mark, and call the first draft in that cross cut Number 1, and so on in succession to the end of it. On his return to the mark where he departed from the main line, let the dialling on the cross cut stand in the book as a *parenthesis*, and let him *resume* his course on the lode, numbering his draft in order from where he branched off. By this system he will have no turning from one place to another in his book — all will be regular; and if the main course, or any other, should be required to be copied separately, in the fair dialling book, it can easily be done. Moreover, should a diagram or geometrical plan of the level and all its windings, and drifts or cross drivings, be required, by this mode of entry everything will appear in its proper place.

Another part of the 'system' is, to let the sight or vane fixed at 360° always take the lead, and the surveyor's eye placed at the opposite vane, except when taking back observations. This will be found under the head of 'Remarks connected with the Converting Table;' and in horizontal dialling, let two drafts be made from every station, which will expedite the work, as the dialler will only have to wait for the settling of the needle once, instead of twice by the other method.

SURVEYING WITHOUT THE MAGNETIC NEEDLE.

This is a valuable modern discovery in mine surveying, and as 'necessity is the mother of invention,' the general introduction of railways and tramroads in mines drove the surveyor to seek some substitute for the needle, which the attraction of iron rendered useless, and he has happily succeeded.

This method of surveying cannot be performed with the common dial; but the best circumferenters are now made with an external graduation and vernier scale, on the theodolite principle, on purpose for the performance of this work.

Mr. W. Cox (from Arnold's, London), of Devonport, makes these instruments in a superior style to any other in the west of England.

The method of surveying, on this principle, differs from the magnetic method chiefly in one particular—namely, that in every fresh draft the position of the bearing must be ascertained by the back observation, in the direction of the sights, and the angle made at the old station must be obtained and preserved at the new station ; and this is evident, because we have no magnet for our guide. For example :—Suppose we are surveying over a railway in a level, and the last observation was 259° ; after measuring the length, the instrument is removed and carried forward to the place of the light where the angle was taken, and a mark and light left at the old station. Then, after the instrument has been adjusted in its true place, the next act of the surveyor is to place the centre of the vernier on 259°, as it stood at the old station ; and if the instrument does not move by rack-work, he must keep all firm with his hands, and turn·the head toward the last station, until the candle is seen through the sights. He then removes behind the instrument, and moves the sights in the direction for the next draft, where the assistant is holding a light for the purpose (the graduation being fixed), and this new draft gives (say) 270¼°, showing a difference between the two drafts of 11¼°. Although this process is somewhat tedious in description, it is simple in practice, and the history of one draft is as well as a hundred ; and we may observe that, with proper care and judgment, this is the most perfect method of surveying, because there is no risk of attraction; and as the circle is much larger than the inside plate, and the divisions more distinct, together with the vernier scale being applied, the angle can be read off to one or two minutes, a nicety which cannot be attained by the needle in the common way. It is hardly necessary to state that, in order to obtain the bearing, there must be at least one draft in the traverse where the needle must be brought into play, and this draft will determine the polarity or direction of the whole.

Further, let it be remarked, that a survey may be resolved into bearings, and worked trigonometrically, when this method is used, as by the needle.

Suppose a case that we are about to survey over a railway, but there is space enough clear of iron for the first draft ; and taking the observation with the needle, we find the north point (a right-hand dial) stand at $176\frac{1}{2}°$; we then fix the outer circle with the vernier precisely at the same point, and then, throwing off the needle, perform all the remainder of the traverse by means of the outer circle. Hence it will be evident, then, if the outward circle is also graduated toward the right hand, that the whole course will come under the immediate operation of the 'converting table,' as if the work had been performed with the needle ; and if the graduation should be reversed, the 'left-hand' bearings will apply accordingly, regard being had to inversion in both cases.

This instrument is also well adapted for taking the bearing of diagonal or underlaying shafts, having a lift of iron pumps—a job that has often baffled the skill and ingenuity of diallers, and occasioned numerous and most serious errors.

The operation may be performed thus: — Suppose we are in the 60 fathom level, and from thence to the 100 the shaft was sunk on the course of the lode, on an underlay of 3 feet per fathom northerly. By applying the instrument at some point in the level near the shaft (but far enough away to be free from attraction by the pumps), we find the bearing by the needle, to a point opposite the shaft, to be due west, and the vernier on the outer rim standing at 90°; we then remove the instrument to the shaft, where the light was held, and adjust the back observation as before directed, having 90° on the outer rim, and the needle thrown off as useless, because we are now close to the pumps. A light is to be carried down the shaft as far as it can be seen, and after the graduated circle has been screwed fast, the rack is applied, and the sights turned until we cut the candle in the bottom of the shaft. This being done, we examine and read off the degree against the point of the vernier, which proves to be (say) $187\frac{3}{4}°$. Now, as when the instrument stood in a due west position the outer circle stood at 90°, and in taking the bearing, it stood at $187\frac{3}{4}°$, therefore, by subtracting 90° from 187$\frac{3}{4}$°, we find the gain to the right hand of west is 97$\frac{3}{4}$°, and, the underlay being northerly, the true bearing of the shaft is 7$\frac{3}{4}$° east of north.

The imperative call for accuracy in cases of this kind will be seen when it is considered that the diagonal part of this shaft is upwards of 40 fathoms, and the underlay 3 feet in a fathom; consequently, the whole base is more than 20 fathoms; and an error in the bearing has the same effect on the survey as if it had been made in taking a horizontal draft of 20 fathoms long, and on which an error of 4° would throw the end of the line nearly 9 feet too far either to the right or left.

Should a dialler be called to do a job of this kind in the absence of a suitable instrument, he may accomplish it in the following manner:—Let him fix a cross-staff in such a position that, through one pair of sights, he can see the candle in the shaft, and in the line of the other pair he has the dial fixed in the level, out of the way of the attraction; consequently, the light in the shaft, and the dial in the level, are two objects forming a right angle with his cross-staff. He then requests his assistant to look at a light held immediately over the head of his cross-staff, through the sights of the dial, and he finds this (say) 12° north of west: and as the bearing of the shaft is exactly at right angles with this line, if the underlay is northerly, the bearing of the shaft will be 12° east of north; if southerly, 12° west of south. The best cross-staffs or instruments, for the express purpose of taking right angles, are now made of a hollow cylindrical shape, of brass, with cuts or apertures for taking the observation ; but a substitute may be used on a pinch, by drawing two lines at right angles on a board, about 6 inches square and an inch thick ; let these lines be cut half an inch deep with a fine saw, and then fix it on a three-feet stand ; if the lines are truly drawn and cut, this rough instrument will serve until a better one can be procured.

CONSTRUCTION.

The old method in laying down a traverse was by drawing a parallel line, and removing the protractor at every draft. The evils of this practice are too glaring to require remark.

Fix your protractor, and lay off as many drafts as will come within the convenient range of your parallel ruler ; number them in order as they stand in your dialling book ; remove the protractor, and lay off the first draft from the centre direct ; then apply the protractor to the centre and No. 2, and make the parallel movement until you touch the end of the last line, or No. 1, and then draw and point off the length of No. 2, and so on through all the drafts you have pointed off from the protractor.

The advantages of laying down or pointing off a number of drafts at one fixing of the protractor, and then applying them in their true length and position, is most conspicuous ; and the geometrician will testify of its superiority, both as it regards accuracy and expedition.

M

CONVERTING TABLE.

Remarks on the following table for converting the degrees recorded in the dialling book of an underground survey into the bearings.

All practical men are aware of the difficulty, hazard, and delay that attend an attempt to obtain the bearing of every draft underground, in a long and complicated survey. The best process is to record the degree, or angle only, at which the needle settles, and after the work is finished underground, then convert the various angles into the real bearing or true direction of each draft, and we may remark, that the bearings *must* be obtained if the work is to be mathematically proved. But as it is not an easy matter to turn a long course of dialling into the bearings, with an assurance of being correct, this table has been constructed for that express purpose; and its utility, simplicity, and perfection have been acknowledged by many practical men.

EXPLANATION.

All circumferentors (dial or miner's compass) are not graduated alike. In all cases, 360° stands at the north point, and 180° at the south; but some are figured toward the right hand, from the north point (which we call a ' right-hand dial '), and others toward the left hand; so that a ' right-hand dial ' has 90° at the east point, and a ' left-hand dial ' has 90° at the west point. This diversity of graduation has often caused much perplexity and confusion among diallers. The following table is contrived to suit both sorts of instruments, and is so plainly arranged and marked as to require but little explanation. It must be specially regarded, that the table has been constructed upon the consideration that the eye of the surveyor has been applied to the south sight or vane standing against 180°; this must be invariably the case. Hence the north sight must always take the lead, and the young practitioner may here be told that in dialling a level and making double, or fore and back drafts, at every station, that although his eye must be placed at the *north* sight, necessarily, for the back observation, yet as the dial has not been turned, the needle will stand to the true degree for the record, and no confusion or liability to error can occur.

In converting an underground survey, or any other, from angles into bearings, it is obviously our first object to know the graduation of the instrument by which the work has been performed ; and if it has been a 'right-hand dial,' and the first draft was 167°, the bearing would be 13° west of south, but if it was done by a 'left-hand dial' the bearing would be 13° east of south. The only thing where a liability to error at all exists in obtaining the bearings by inspection from this table, and where caution is required, is in applying the fractions of degrees when they occur in the drafts. On these occasions, observe that when the angle and bearing progress alike, as in all the left-hand side of the column, then the fraction must be *added* to the whole number of the bearing, but otherwise, as in the right-hand side, the fraction must be *deducted* from the whole number.   Lastly, the following desirable proof may be resorted to :—If the course has been correctly converted, the degree and bearing added together or subtracted from each other will make one of the following numbers :—0, 90, 180, 270, 360 ; and this may be done almost at a glance after the survey has been converted into bearings.

TABLE

FOR CONVERTING ANGLES INTO BEARINGS.

Rt.Hd. Dial W. of N. / Lt. Hd. Dial E. of N.		Rt. Hd. Dial N. of W. / Lt. Hd. Dial N. of E.		Rt. Hd. Dial S. of W. / Lt. Hd. Dial S. of E.		Rt. Hd. Dial W. of S. / Lt. Hd. Dial E. of S.	
Angle	Bearing	Angle	Bearing	Angle	Bearing	Angle	Bearing
1 is	1	46 is	44	91 is	1	136 is	44
2	2	47	43	92	2	137	43
3	3	48	42	93	3	138	42
4	4	49	41	94	4	139	41
5	5	50	40	95	5	140	40
6	6	51	39	96	6	141	39
7	7	52	38	97	7	142	38
8	8	53	37	98	8	143	37
9	9	54	36	99	9	144	36
10	10	55	35	100	10	145	35
11	11	56	34	101	11	146	34
12	12	57	33	102	12	147	33
13	13	58	32	103	13	148	32
14	14	59	31	104	14	149	31
15	15	60	30	105	15	150	30
16	16	61	29	106	16	151	29
17	17	62	28	107	17	152	28
18	18	63	27	108	18	153	27
19	19	64	26	109	19	154	26
20	20	65	25	110	20	155	25
21	21	66	24	111	21	156	24
22	22	67	23	112	22	157	23
23	23	68	22	113	23	158	22
24	24	69	21	114	24	159	21
25	25	70	20	115	25	160	20
26	26	71	19	116	26	161	19
27	27	72	18	117	27	162	18
28	28	73	17	118	28	163	17
29	29	74	16	119	29	164	16
30	30	75	15	120	30	165	15
31	31	76	14	121	31	166	14
32	32	77	13	122	32	167	13
33	33	78	12	123	33	168	12
34	34	79	11	124	34	169	11
35	35	80	10	125	35	170	10
36	36	81	9	126	36	171	9
37	37	82	8	127	37	172	8
38	38	83	7	128	38	173	7
39	39	84	6	129	39	174	6
40	40	85	5	130	40	175	5
41	41	86	4	131	41	176	4
42	42	87	3	132	42	177	3
43	43	88	2	133	43	178	2
44	44	89	1	134	44	179	1
45	45	90 { R. H. D. W. / Lt. H. D. E. }	135	45	180 South	

TABLE
FOR CONVERTING ANGLES INTO BEARINGS.

Rt. Hd. Dial E. of S. Lt. Hd. Dial W. of S.		Rt. Hd. Dial S. of E. Lt. Hd. Dial S. of W.		Rt. Hd. Dial N. of E. Lt. Hd. Dial N. of W.		Rt. Hd. Dial E. of N. Lt. Hd. Dial W. of N.	
Angle	Bearing	Angle	Bearing	Angle	Bearing	Angle	Bearing
181	is 1	226	is 44	271	is 1	316	is 44
182	2	227	43	272	2	317	43
183	3	228	42	273	3	318	42
184	4	229	41	274	4	319	41
185	5	230	40	275	5	320	40
186	6	231	39	276	6	321	39
187	7	232	38	277	7	322	38
188	8	233	37	278	8	323	37
189	9	234	36	279	9	324	36
190	10	235	35	280	10	325	35
191	11	236	34	281	11	326	34
192	12	237	33	282	12	327	33
193	13	238	32	283	13	328	32
194	14	239	31	284	14	329	31
195	15	240	30	285	15	330	30
196	16	241	29	286	16	331	29
197	17	242	28	287	17	332	28
198	18	243	27	288	18	333	27
199	19	244	26	289	19	334	26
200	20	245	25	290	20	335	25
201	21	246	24	291	21	236	24
202	22	247	23	292	22	337	23
203	23	248	22	293	23	338	22
204	24	249	21	294	24	339	21
205	25	250	20	295	25	340	20
206	26	251	19	296	26	341	19
207	27	252	18	297	27	342	18
208	28	253	17	298	28	343	17
209	29	254	16	299	29	344	16
210	30	255	15	300	30	345	15
211	31	256	14	301	31	346	14
212	32	257	13	302	32	347	13
213	33	258	12	303	33	348	12
214	34	259	11	304	34	349	11
215	35	260	10	305	35	350	10
216	36	261	9	306	36	351	9
217	37	262	8	307	37	352	8
218	38	263	7	308	38	353	7
219	39	264	6	309	39	354	6
220	40	265	5	310	40	355	5
221	41	266	4	811	41	856	4
222	42	267	3	312	42	357	3
223	43	268	2	313	43	358	2
224	44	269	1	314	44	359	1
225	45	270 { R.H.D.E. / L.H.D.W. }		315	45	360 North.	

APPLICATION OF THE CONVERTING TABLE.

Suppose the needle stood at $246\frac{1}{2}°$ what is the bearing?

Answer $\begin{cases} \text{By a right-hand dial } 23\frac{1}{2}° \text{ South of E.} \\ \text{By a left-hand dial } 23\frac{1}{2}° \text{ South of W.} \end{cases}$

It may be remarked that the table is equally applicable for changing bearings into angles if required. For example : — An observation was made with a right-hand dial, and the bearing found to be 27° 17′ E. of N. : at what degree did the needle point?

Ans. 332° 43′, and if proof is required it will be seen that the sum of these degrees and minutes is 360°.

EXAMPLE.

Convert the following angles taken with a left-hand dial into bearings :—

$210\frac{1}{4}°$ $176\frac{1}{2}°$ $305\frac{3}{4}°$ $28\frac{1}{2}°$ $107\frac{1}{8}°$ $97\frac{3}{4}°$

OPERATION.			PROOF.
$210\frac{1}{4}°$ is $30\frac{1}{4}°$	W. of S.	$210\frac{1}{4}°-30\frac{1}{4}°=180°$
$176\frac{1}{2}$ $3\frac{1}{2}$	E. of S.	$176\frac{1}{2} + 3\frac{1}{2} =180$
$305\frac{3}{4}$ $35\frac{3}{4}$	N. of W.	$305\frac{3}{4} -35\frac{3}{4} =270$
$28\frac{1}{2}$ $28\frac{1}{2}$	E. of N.	$28\frac{1}{2} -28\frac{1}{2} = 0$
$107\frac{1}{8}$ $17\frac{1}{8}$	S. of E.	$107\frac{1}{8} -17\frac{1}{8} = 90$
$97\frac{3}{4}$ $7\frac{3}{4}$	S. of E.	$97\frac{3}{4} - 7\frac{3}{4} = 90$
348 12	W. of N.	$348 +12 =360$

Convert the following angles taken with a right-hand dial into bearings :—

9° 45'   239° 25'   331° 12'   160° 58'   45° 6'

OPERATION.			PROOF.		
9° 45' is 9° 45' W. of N.			9° 45' − 9° 45' = 0°		
239 25	30 35	S. of E.	239 25 + 30	35	= 270
331 12	28 48	E. of N.	331 12 + 28	48	= 360
160 58	19 2	W. of S.	160 58 + 19	2	= 180
45 6	44 54	W. of N.	44 54 + 45	6	= 90

N.B.—In practice it would not be necessary or convenient to *state* proofs—it is introduced here for the learner's sake, that he may be enabled to insure certainty in this essential matter.

In pressing on our young mining friends the advantage of adopting a perfect system, we advise that in preparing a course of dialling for trigo-metrical solution, by changing the angles into bearings, care should be taken that all the drafts should be made either to exceed 45°, or that they should all stand below, or at least not exceed that half quadrant. Our reason for being urgent on this matter is, that there may be a uniformity in placing the sides in the traverse table after the draft has been computed. And let it be par-ticularly noticed that, if the bearings are not suf-fered to exceed 45°, that the *last* expression of the bearing will signify the *longer* of the two sides. That is, suppose a draft taken underground was 287¼°, measuring 45 feet 8 inches ; now looking at the 'converting table' we see that, if this draft was taken with a 'left-hand dial,' the bearing is 17¼° nor of west (or N. of W.), and the two sides will be found by computation to be 13 feet 7 inches, and 43 feet 7 inches. Query, into what columns respectively must these numbers be placed? As the bearing was north of west, and our system states that 'the last expression of the bearing will signify the longer of the two sides,' consequently the longer side (43 feet 7 inches) must be placed in the '*west*' column, and 13 feet 7 inches in the north column.

If this order is followed up, it will render the working of traverses (which is the most important operation in mine surveying) a plain, pleasing, and satisfactory exercise. In this edition we would needs bring forward everything likely to promote the advancement of the young mining officer in this paramount branch of his profession, and therefore give him to understand that, in traversing, there must be a regular course from beginning to end.

We shall make ourselves understood in this matter, by taking a case where a person makes a survey for the purpose of ascertaining the length and bearing of a level driven on an east and west lode; and, for some convenient purpose, he begins his dialling at some point about the middle of the level, and dials from thence to the eastern end; he then returns to the station or start at the middle of the level, and dials on to the western end, and thus completes the survey.

Now if he were to proceed to work the traverse from his dialling book in this state, his results would appear as if his level were almost without length or bearing, as his eastings would be balanced by his westings, &c.

In order to go *systematically* to work in this case, his first operation must be to *reverse* the order of one or the other of the diallings; that is, if he pleases to let the first remain, which is the eastern dialling, and would accommodate the western part to suit the other, he must alter or reverse all the drafts by converting (say) 16° south of west into 16° north of east, and so of all the rest.

In winding up this course of instruction, we will take a short survey, and go through with it at length, and the student may accompany us if he pleases; for we are still of the same opinion as when we wrote the first volume, that practical teaching is the best.

EXAMPLE.

It is required to sink a vertical shaft on the end of a level, and the diallings from the bottom of an old downright shaft are as follows :—

Surveyed with a 'right-hand' dial.

			fath.	ft.	in.
No. 1.	356¼°	Length	18	3	0
2.	84½		12	1	6
3.	98		15	4	0

fath. ft.
4. A Winze 322° Underlay 25¼°, Inclined Length 11 2
5.   107¾° Length.   25 5 6 End.

This is the underground work, and our first

operation is to find out the underlay of the winze, in order that it may stand as a common draft in the survey.

OPERATION.

The underlay, or angle made by the dip of the winze and a vertical line, being 25½ degrees, we find it standing in the first table against 2 feet 7 inches, showing that every fathom of the winze gives a base of 2 feet 7 inches, and the length of the winze being 11 fathoms 2 feet, we multiply

$$
\begin{array}{r}
\text{ft. in.} \\
\tfrac{1}{3}) \quad 2 \ \ 7 \\
11{\cdot}2 \\
\hline
4 \ 4 \ 5 \\
10 \\
\hline
4 \ 5 \ 3 \\
\hline
\end{array}
$$

fath. ft. in.

Here we find the base of the winze to be 4 5 3

The next thing is to refer to the converting table to reduce the drafts into bearings; taking special notice that the work was done with a right-hand dial.

We therefore find that No, 1. 356¼° is 3¾° E. of N.

2. 84½ - 5½ N. of E.

3. 98 - 8 S. of W.

Winze 4. 322 - 28 E. of N.

5. 107¾ - 17¾ S. of E.

Our work is now prepared for entry in the traverse table as data for trigonometrical computation.

No.	Angles and Lines		Trigonometrical Results			
Draft	Bearings	Lengths	East	West	North	South
		fath. ft.in.				
1	3¾° E. of N.	18 3 0				
2	5¼ N. of E.	12 1 6				
3	8 S. of W.	15 4 0				
4	28 E. of N.	4 5 3				
5	17¾ S. of E.	25 5 6				

The above is the table with the bearings and lengths of the drafts entered in order for receiving the trigonometrical results in their proper and respective columns, and that everything may be clear to the learner we shall let this table remain as it is, and make a similar one, in which the computations are entered, and proceed to take out the tabular numbers from the first mathematical table, and multiply them by their respective lengths.

FIRST DRAFT.

$$∠ 3¾° \text{ Tabular}$$

ft. in.	ft. in.
0 4·7	Tabular 5 11·85
6	6
2 2·2	35 11·10
3	3
6 6·6	107 9·30
2·3	2 11·92
6 8·9 Easting.	110 9·2 Northing.

Now the sides of the triangle formed by the first draft are ready to be transferred to the east and north columns of the traverse table.

SECOND DRAFT.

	ft. in.		ft. in.
∠ 5¼° Tabular	0 6·9	Tabular	5 11·67
	12¼		12¼
	6 10·8		71 7·92
	1·7		1 11·42
	7 0·5 Northing.		73 7·3 Easting.

When the bearing does not diverge much from the cardinal point, there is but little difference between the length of the hypothenuse and the longest of the legs, as in the right-hand sides of the above two drafts.

THIRD DRAFT.

	ft. in.		ft. in.
∠ 8° Tabular	0 10·02	Tabular	5 11·3
	8		8
	6 8·16		47 6·4
	2		2
	13 4·32		95 0·8
	3·31		1 11·8
	13 1·0		93 1·0 Westing.

The length of the draft being 15 fathoms 4 feet, we have multiplied by 16, and deducted ⅓ as the shortest method.

FOURTH DRAFT, OR BASE OF WINZE.

	ft. in. in.		ft. in. in.
∠ 28° Tabular	2 9·8 or 31·8	Tabular	5 3·6 or 63·6
	5		5
	169·0		318·0
	4·2		7·9
	12)164·8		12)310·1
	13·8·8 Easting.		25·10·1 Northing.

In the above, it will be seen that we have thrown the tabular length into *inches* and parts, and the practitioner will find this, in general, the easiest way of calculating.

FIFTH DRAFT.

	ft. in	in.		ft. in.	in.
$\angle 17\frac{3}{4}°$ Tabular	1 10·0	or 22·0	Tabular 5	8·6	or 68·6
		26			26
		132			411·6
		44			1372
		572·0			1783·6
		1·8			5·7
		12)570·2			12)1777·9
		47·6·2 Southing.			148·1·9

Now the computations are ready for entry in the following table:—

No.	ANGLES AND LINES		TRIGONOMETRICAL RESULTS			
Draft	Bearings	Lengths	East	West	North	South
		fath. ft. in.	ft. in.	ft. in.	ft. in.	ft. in.
1	3¾° E. of N.	18 3 0	6 8·9	- -	110 9·2	
2	5¼ N. of E.	12 1 6	73 7·3	- -	7 0·5	
3	8 S. of W.	15 4 0	- -	93 1·0	- -	13 1·0
4	28 E. of N.	4 5 3	13 8·8	- -	25 10·1	
5	17¾ S. of E.	25 5 6	148 1·9	- -	- -	47 6·2
			242 2·9	93 1·0	143 7·8	60 7·2
			93 1·0		60 7.2	
			149 1·9	Easting.	83 0·6	Northing

Now we might proceed to lay down the position or place of our new vertical shaft at the surface without any further operation. For by measuring off from the centre of the old shaft at surface 149 feet 2 inches, due east, and from the end of that line measuring 83 feet, due north, would bring us exactly over the end of the fifth or last draft, where the shaft is to come down, but we would work out the direct length and bearing also, as before described, and apply it.

PROBLEM.

At the celebrated South Caradon Copper Mine a new vertical shaft was commenced in the early part of 1842, which is intended to intersect the main lode at the depth of 100 fathoms below the adit level, which is about 40 fathoms from surface in the vicinity of the new shaft. From a point in this level, a drift or cross-cut has been begun, and designed to be driven in a direct line to the centre of the new shaft, and from thence to rise against it, if necessary : and the aim and object of the survey is to ascertain the exact length and bearing of the said cross-cut, as every proper means have been adopted to certify that it has been commenced at the nearest point to the shaft.

The following is the whole course of dialling in its most simplified form, with the irregular surface lines reduced to horizontal measure, the angle of every draft converted into bearing, and the whole given in complete order for working the traverse without any preliminary preparation. The draft standing on the top of 934 feet is from the centre of the new shaft at surface to a line hung in the old engine shaft, which is also vertical to the adit, and the next draft is taken from that line in the adit and continued to the end of the 34th draft,

through the same level, where the cross-cut com-
mences.

It is also required to furnish the *bearing* of the
lode from the 3rd to the 34th draft inclusive.

REMARKS.

As this course of dialling has been rendered so
plain, there appears to be no occasion to introduce a
double entry of it, as the field and underground
work is sufficiently manifest in the first three
columns of the following table, in combination with
the trigonometrical operation. It may be satisfactory to the student to be informed that this work
has been accomplished and proved to be perfectly
correct. The cross-cut was driven all the way (50
fathoms) through a hard granite country, at nearly
1000*l.* cost, and occupied full two years of uninterrupted labour.

COMPUTATION.

No.	Angles and Lines		Trigonometrical Results			
Drafts	Bearings	Lengths	East	West	North	South
	From new Shaft to Engine Shaft.	ft. in.	ft. in.	ft. in.	ft. in.	ft. in.
	14° 18′ S. of W.	934 0	- -	905 0·7	- -	230 8·3
1	13¾ S. of E.	30 0	29 0·7	- -	- -	7 1·6
2	4¾ E. of S.	30 6	2 6·3	- -	- -	30 4·8
	On course of lode.					
3	5¼ S. of E.	28 6	28 4·5	- -	- -	2 8·8
4	4 N. of E.	42 6	42 4·7	- -	2 11·6	
5	16 S. of E.	30 8	29 5·8	- -	- -	8 5·4
6	2½ do.	32 0	31 11·7	- -	- -	1 4·7
7	18¼ do.	29 2	27 8·9	- -	- -	9 1·6
8	4½ do.	18 0	17 11·5	- -	- -	1 4·0
9	3¼ N. of E.	36 0	35 10·2	- -	2 2·4	
10	13 S. of E.	17 0	16 6·8	- -	- -	3 9·9
11	19¾ do.	26 6	24 11·2	- -	- -	8 11·4
12	14 do.	22 7	21 11·0	- -	- -	5 5·6
13	15 do.	36 3	35 0·1	- -	- -	9 4·6
14	1 do.	46 6	46 5·9	- -	- -	0 9·8
15	17 N. of E.	41 3	39 5·4	- -	12 0·7	
16	4 do.	11 2	11 1·7	- -	0 9·3	
17	21½ do.	12 8	11 9·4	- -	4 7·7	
18	1 do.	27 0	27 0·0	- -	0 5·7	
19	2 do.	38 8	38 7·7	- -	1 4·2	
20	1 S. of E.	18 0	18 0·0	- -	- -	0 3·8
21	20 do.	12 6	11 8·9	- -	- -	4 3·4
22	7 do.	11 6	11 5·0	- -	- -	1 4·8
23	1 do.	25 8	25 8·0	- -	- -	0 5·3
24	4 N. of E.	31 6	31 5·1	- -	2 2·3	
25	13 S. of E.	34 6	33 7·4	- -	- -	7 9·1
26	1 N. of E.	18 8	18 8·0	- -	0 3·9	
27	5 do.	28 0	27 10·7	- -	2 5·3	
28	8 S. of E.	65 0	64 4·4	- -	- -	9 0·6
29	2 do.	36 6	36 0·4	- -	- -	1 3·3
30	5 do.	18 0	17 11·2	- -	- -	1 6·8
31	6½ N. of E.	24 8	24 6·0	- -	2 9·5	
32	1 do.	12 3	12 3·0	- -	0 2·6	
33	18 do.	29 6	28 0·7	- -	9 1·4	
34	12 do.	20 1	19 7·8	- -	4 2·1	
			899 6·1	905 0·7	45 8·7	345 9·6
				899 6·1		45 8·7
			Westing	5 4·6	Southing	300 1·9

N

Then by the foregoing method of proceeding, the required answers will be found as follows :—

Length of cross-cut from adit to centre of shaft, 300 feet 2 inches.

Bearing of cross-cut 1° 2′ west of north.

Bearing of lode from 3rd to 34th draft, inclusive, 2⅛° south of east.

PROBLEM.

A lode was opened on the back by costeening in several places, and its course, by compass, found to be 10½° south of east ; but this was on the ascent of a steep hill whose angle of elevation was 16½°, and the lode underlaying northerly three feet in a perpendicular fathom.

Query. What is the true bearing or course of the lode? and what would be the amount of error in carrying on the line 600 fathoms (horizontal measure), supposing the run of the back of the lode on the ascent had been taken, by mistake, instead of the true horizontal course?

OPERATION.

We find in the first table that 16½° of elevation gives 1 ft. 8·45 in. perpendicular, and 5 ft. 9·04 in. base for the corresponding sides of the triangle.

Hence If 6 perp. gives 3 underlay, what will 1 8·45 give ?

ft.	ft.	ft. in.
12	12	12
72	36	20·45
		36
		72)736·20(10·2

Showing that the underlay of the lode carries the line 10¼ inches further north than the line taken at the surface (or bearing) on every horizontal line of 5 ft. 9·04 in. Therefore we have the two sides of a right-angled triangle 5 ft. 9·04 in. and 10·2 in., and the angle opposite the shortest side will be the amount of the angle of error.

```
        ft.   in.       in.     ft.
  As 5  9·04    :   10·2  ::  6
       12                    12
      ─────                 ─────
      69·04                  72
                           10·2 in.
              69·04)7 34·4(10·63
```

By inspection of the 2nd table we find the nearest next less angle standing opposite this number (10·63 in.) is 8° 15′, giving 10·44 in.; and as the difference between this and the next greater (8½) is ·32, and the difference between 10·63 in. and 10·44 in. is ·19, we say,

As ·32 is to ·19, so is 15′ = 9′.

And 9 added to 8° 15′ gives 8° 24′ for the angle of error; and by deducting this angle of error (if it may be so called) from the course of the lode on the inclined surface (10½ south of east), we have 2° 6′ south of east for the true bearing of the lode; and as the error is 10·64 in. in a horizontal fathom, this number multiplied by 600 gives 6384·00 inches, or 88⅔ fathoms, too far south, on a line of 600 fathoms.

We beg to call the particular attention of our young mine agents to this case. The vast sums of money expended every year in this country in costeening or right-angle open cuttings at the surface in searching for lodes, is well known to practical men ; and probably there is no branch of mining in which there is a greater waste of time and money. Workmen are often put to co-steen at random, when a scientific survey might have put them in a position for opening the lode in a few hours which has occupied them for weeks or months, and at the last all their labours fraught with doubt and uncertainty ; but these are not the worst consequences arising from ignorance or in-attention to this subject. I know a rich mine in this county, where the angle of ascent was made in error, and carried on to find the lode in the ad-joining set, and the effect was, that they pitched to cut the lode in a new set nearly 200 fathoms out of the line ; and they have been driving now nearly five years through a hard country at a cost of some thousands of pounds, and have not yet cut the lode ! Volumes might· be written on the errors that have taken place from this source ; and probably there is not a man of experience to be found but what can confirm the report with his testimony.   ·

PROBLEM.

In a 20 fathom level driving on an east and west lode, underlaying north, a winze has been commenced bearing due north, and it is determined to pitch a rise against it in the 40 fathom level (the 30 fathom level not having been driven far enough east to rise from). The following is a statement of the dialling from the middle of the above winze, in the 20, through the level west towards another winze sunk to the 30 fathom level :—

No.		ft.	in.	No.		ft.	in.
1	$282\frac{1}{4}°$	59	10	3	$264\frac{1}{2}°$	33	4
2	$286°$	61	8	4	$260\frac{1}{4}°$	77	3

This brings us to the brace of the winze communicating with the 30 fathom level, which we may call No. 5,—its diagonal length, 65 feet ; underlay, $22\frac{1}{8}$ degrees ; bearing, $9\frac{1}{4}°$ east of north. From the foot of this winze in the 30, the dialling is continued westerly to another winze communicating with the 40 fathom level ; viz. :—

No.		ft.	in.
6	$272\frac{1}{2}°$	60	9
7	$256°$	52	0
8	$287\frac{1}{4}°$	45	8

We now arrive at the brace of the winze to the 40, which we call No. 9,—length, 70 ft. 6 in. ; underlay, $31\frac{1}{2}°$; bearing, 4° west of north. From the bottom of this winze in the 40, the course turns easterly, and is continued in that direction ; viz. :—

No.		ft.	in.	No.		ft.	in.
10	83°	85	4	13	90°	77	6
11	77°	28	5	14	$92\frac{1}{2}°$	23	8
12	$104\frac{1}{4}°$	76	0	15	$99°$	107	2

At the end of this 15th draft we place an assumed mark in the back of the 40 fathom level.

Question. It is requested to know how far we must measure east or west from this mark in order to arrive at the exact point for rising against the winze sinking from the 20 fathom level? also what will be the average underlay at that place, and what will be the length of the winze from the 20 to the 40 fathom level?

COMPUTATION.

(SURVEYED WITH A LEFT-HAND DIAL.)

No.	Angles and Lines			Trigonometrical Results			
Draft	Degree	Bearing	Length	East	West	North	South
			ft. in.	ft. in.	ft. in.	ft. in.	ft. in.
1	282¼	12¼° N. of W.	59 10	- -	58 6	12 8	
2	286	16 N. of W.	61 8	- -	59 3	17 0	
3	264½	5½ S. of W.	33 4	- -	33 2	- -	3 2
4	260¼	9¾ S. of W.	77 3	- -	76 2	- -	13 1
5	Winze	9¼ E. of N.	24 10	4 0	- -	24 6	
6	272½	2½ N. of W.	60 9	- -	60 8	2 8	
7	256	14 S. of W.	52 0	- -	50 6	- -	12 7
8	287¼	17¼ N. of W.	45 8	- -	43 7	13 7	
9	Winze	4 W. of N.	36 10	- -	2 7	36 9	
10	83	7 N. of E.	85 4	84 8	- -	10 4	
11	77	13 N. of E.	28 5	27 8	- -	6 5	
12	104¼	14¼ S. of E.	76 0	73 8	- -	- -	18 9
13	90	East	77 6	77 6			
14	92½	2½ S. of E.	23 8	23 8	- -	- -	1 0
15	99	9 S. of E.	107 2	105 10	- -	- -	16 9
				397 0	384 5	123 11	65 4
				384 5		65 4	
			Easting	12 7	Northing	58 7	

Now, we discover by the foregoing calculation that our assumed mark in the 40 fathom level is 12 feet 7 inches *too far east* for the central point of the rise against the winze sinking from the 20, and which was the paramount object required.

NOTE.—As the last draft (reversed) is 9° N. of W., in order to be quite accurate it will require us to measure 12 feet 9 inches back through the level to make good 12 feet 7 inches westing, and this will give 1 foot 10 inches more of base or northing, which added to 58 feet 7 inches, the underlay shown by the column of northing, will give the answer to the question, 60 feet 5 inches for the whole underlay.

The first operation in working this problem is to find the base and perpendicular of the two winzes numbers 5 and 9, and their respective bases form the operative lines in the above traverse.

ft. in.   ft. in.

Winze No. 5 gave 24 10 base, and 60 1 perpendicular.

Winze No. 9 gave 36 10 base, and 60 1 perpendicular.

Now the vertical depth being 120 2, and the base or northing 60 feet 5 inches, we have thus the two sides of a right-angled triangle to find the hypothenuse and dip or angle of declination, which on trial will be found—

Hypothenuse, or length of winze, 134 feet.

Angle, or underlay of winze, 26¼ degrees.

PROBLEM.

A tunnel has been commenced at the foot of a hill, and is intended to be driven through it.

The bearing from the above point, or the course of the tunnel, is to be due east, and it is required to know the exact corresponding point on the other side of the hill, in order to set another company of men to drive a *dead* level to meet the drivings that are progressing from the west side.

The length of the tunnel is also required.

The following is the survey from the first point:—

ft.

	No.			Length	
	1	Elevation	14°	Length	26
	2	—	12¼	—	26
	3	—	11	—	17
	4	—	18½	—	90
	5	—	10	—	60
	6	—	7⅛	—	119
	7	Horizontal	0	—	29
	8	Depression	5¾	—	28
	9	—	16	—	230

Judging that we have now arrived somewhere near the level or horizontal plane of the start, or that our ' depressions' have made good our ' elevations,' we place an assumed mark at the end of the last or 9th draft, and retire to work out our lines and angles by trigonometry.

OPERATION.

	PERP.	BASE.
fath. ft.	ft. in.	ft in.
No. 1 Elevation 14° Length 4 2 Tabulars 1 5·4	5 9·9
4⅓	4⅓
5 9·6	23 3·6
5·8	1 11·3
6 3·4	25 2·9

Thus we find the 1st draft gives a rise or elevation of 6 feet 3·4 inches, and base or horizontal length 25 feet 2·9 inches ; and proceeding in the same manner with all the drafts, and finding the difference between the elevations and depressions,

we shall obtain true data for correcting our assumed
mark, and replacing it in its proper position.

		Elevation			Horizontal	
		ft.	in.		ft.	in.
No. 1	gives	6	3·4	and	25	2·9
2		5	6·2		25	4·9
3		3	2·9		16	8·3
4		28	6·0		85	4·0
5		10	5·0		59	0·1
6		14	8·5		118	1·0
		68	8·0			
7					29	0·0
		Depression				
8	gives	2	8·2		27	10·5
9		63	3·8		220	3·0
		66	0·0		606	10·7

Now as the depressions are 2 feet 8 inches less
than the elevations, it demonstrates that our assumed mark is 2 feet 8 inches too high, and as the
declination of the ground from the last draft east-
ward continues on the same angle of depression of
16 degrees, we have perpendicular 2 feet 8 inches
and angle 16° to find the corresponding hypothenuse and base ; and by inspection of the 2nd table
we see that the 'tabulars' opposite 16° are 1 foot
8·6 inches, and 6 feet 2·9 inches hypothenuse

Therefore, if 1 foot 8·6 inches give 6 feet 2·9
inches, what will 2 feet 8 inches give?

Which will be found to give 9 feet 8 inches of
hypothenuse.

And by the 1st table it will be found that 9 feet

8 inches of hypothenuse on an angle of 16° will give for the longest side, or base, 9 feet 4 inches.

ADJUSTMENT.

By removing the assumed mark 9 feet 8 inches, due east on the slope, we fix on the exact spot for commencing the eastern end of the tunnel, and we need hardly observe, that the two extreme marks mean the bottom or floor of the tunnel.

Then by adding the base, 9 feet 4 inches, made by the corrections, to the sum of the horizontals, 606 feet 10·7 inches, we have just 616 feet 3 inches for the length of the tunnel.

NOTE.—Should it be required to put down vertical shafts on the tunnel, the foregoing computations reveal what their depths would be respectively at all parts of the tunnel, and the deepest shaft would be 11 fathoms 2 feet 8 inches at the end of the 6th draft, and 55 fathoms from the western mouth of the tunnel.

PROBLEM.

It is intended to sink a shaft on the end of a level driven from Pendarves' shaft, and the following is the survey from the centre of Pendarves' shaft to the end of the level, viz :—

No.			ft. in.	No.			ft. in.
1	3°	W. of N.	45 0	6	$6\frac{3}{4}$°	S. of E.	27 0
2	$7\frac{1}{2}$	N. of E.	24 6	7	15	S. of E.	16 5
3	$8\frac{1}{2}$	N. of E.	18 0	8	5	N. of E.	21 0
4		East	49 1	9	$12\frac{1}{4}$	N. of E.	14 7
5	12	S. of E.	30 0	10	9	W. of N.	28 0

As profound accuracy is required in this case (it being intended to facilitate the work by rising against the new shaft from the end of the level), a reverse or proof course of dialling -has been made from the end back to the centre of Pendarves' shaft; viz. :—

No.		ft.	in.	No.		ft.	in.
1	$8\frac{1}{2}°$ E. of S.	26	10	6	$1\frac{1}{4}°$ S. of W.	44	0
2	11 S. of W.	15	0	7	7 S. of W.	26	0
3	$4\frac{3}{4}$ S. of W.	19	6	8	$9\frac{3}{4}$ S. of W.	22	8
4	$13\frac{1}{4}$ N. of W.	20	0	9	$2\frac{1}{4}$ E. of S.	43	10
5	$9\frac{1}{2}$ N. of W.	52	3				

It is now required to know if there is an exact agreement between these two surveys, or fore and back diallings (or what is the difference between them), and if so, what is the length and bearing from the centre of Pendarves' shaft, at the surface, to the point exactly over the end of the level where the centre of the new shaft must be fixed?

OPERATION.

FROM PENDARVES' SHAFT TO EASTERN END.

No.	Angles and Lines		Trigonometrical Results			
Drafts	Bearings	Lengths	East	West	North	South
		ft. in.	ft. in.	ft. in.	ft. in.	ft. in.
1	3° W. of N.	45 0	- -	2 4·2	44 11·2	
2	$7\frac{1}{4}$ N. of E.	24 6	24 3·7	- -	3 1·1	
3	$8\frac{1}{2}$ N. of E.	18 0	17 9·6	- -	2 7·9	
4	East	49 1	49 1·0			
5	12 S. of E.	30 0	29 4·1	- -	- -	6 2·8
6	$6\frac{3}{4}$ S. of E.	27 0	26 9·8	- -	- -	3 2·1
7	15 S. of E.	16 5	15 10·3	- -	- -	4 3·0
8	5 N. of E.	21 0	20 11·0	- -	1 10·0	
9	$12\frac{1}{4}$ N. of E.	14 7	14 3·0	- -	3 1·1	
10	9 W. of N.	28 0	- -	4 4·6	27 7·9	
			198 4·5	6 8·8	83 3·2	13 7·9
			6 8·8		13 7·9	
		Easting	191 7·7	Northing	69 7·3	

FROM EASTERN END TO PENDARVES' SHAFT.

No.	ANGLES AND LINES		TRIGONOMETRICAL RESULTS			
Drafts	Bearings	Lengths	East	West	North	South
		ft. in.	ft. in.	ft. in.	ft. in.	ft. in.
1	8½° S. of E.	26 10	3 11·6	- -	- -	26 6·5
2	11 S. of W.	15 0	- -	14 8·7	- -	2 10·3
3	4¾ S. of W.	19 6	- -	19 5·2	- -	1 7·4
4	13¼ N. of W.	20 0	- -	19 5·6	4 7·0	
5	9½ N. of W.	52 3	- -	51 6·5	8 7·5	
6	1¼ S. of W.	44 0	- -	43 11·9	- -	0 11·5
7	7 S. of W.	26 0	- -	25 9·7	- -	3 2·0
8	9¾ S. of W.	22 8	- -	22 4·1	- -	3 10·1
9	2¼ E. of S.	43 10	1 8·6	- -	- -	43 9·6
			5 8·2	197 3·7	13 2·5	82 9·4
				5 8·2		13 2·5
			Westing	191 7·5	Southing	69 6·9

Now we find that as the westing and southing of
the back dialling corresponds with the easting and
northing of the direct dialling to the fraction of an
inch, it amounts to a mathematical demonstration
of the perfection óf the underground survey. It
now only remains for us to obtain the hypothenuse
and angle opposite the base of the two given sides
of the triangle formed by the easting 191 feet 7¾
inches, and northing 69 feet 7 inches, which will be
found to give—

Length (from centre of Pendarves' shaft to point
over end), 203 feet 11¼ inches.

Bearing 20 degrees north of east.

N.B.—After this length and bearing has been
applied at the surface, and the point fixed for the
centre of the new shaft, an infallible and desirable
proof that this last and important work has been

done correctly may be obtained by availing ourselves of the ready means placed within our reach by the cardinal points or sides of the great triangle. Thus, by measuring off from the centre of Pendarves' shaft 191 feet $7\frac{3}{4}$ inches due east, and then from the end of that line 69 feet 7 inches due north, the end will fall exactly on the point fixed at the end of the line 203 feet $11\frac{1}{4}$ inches on the bearing of 20 degrees north of east, if the whole has been done correctly.

PLANS AND SECTIONS OF MINES.

—◆—

PERSONS who have not had practical experience in mining often acknowledge that they find great diffi- culty in comprehending the plans and sections of a mine, or of having a true idea of the workings from an inspection of the drawings. This obscurity may be occasioned from an imperfection in the plans; for if they have been executed under a good *system*, it can hardly fail to exhibit clearly every part of the workings, and, indeed, if the diagrams have not been executed perfectly, and according to rule and order, even miners themselves cannot comprehend them. It requires four distinct ma- thematical or geometrical drawings to represent a mine, which we will briefly notice under each head; and we may observe, as we pass on, that the common cause of people in general not understand- ing the plans is because they expect to know too much from one single drawing. Every separate plan exhibits both a true and false view of some parts of the mine; and the knowledge necessary for the observer is, what parts of the workings it is that each drawing furnishes a true delineation of.

With this introduction we proceed to state that the set of drawings may be described thus:—

1. Ground plan.
2. Horizontal or working plan.
3. Longitudinal section.
4. Transverse section.

And taking them in the order in which they have been placed, we begin with—

(1.) THE GROUND PLAN.

This is, in the main, a general survey of the whole set, or land granted to the adventurers for the purpose of mining. This plan may be on a scale of three or four chains to an inch; and every lord or landowner's bounds should be distinctly marked on this map.

All the lodes are laid down with their true position and course on this plan, as far as they can be ascertained; and we may remark, that this survey should be made at the outset or plant of a mine, and before anything has been determined as to the position of an engine shaft, or any other important work, so that the manager may have the benefit of this map, with the lodes, cross courses, and every necessary thing faithfully delineated thereon, to assist his judgment in forming the most judicious arrangement for future operations. For want of this precaution, how often is it that shafts have been sunk in improper places, to the endless disadvantage of the company; and sometimes they have been abandoned and new shafts sunk, at a fearful loss of time and money! In fact, we believe there are but few mines where the conductors have not had cause to regret;

ultimately, that they had not taken another position for sinking the principal shafts, and which might have been known, at the outset, if the necessary steps had been pursued.

On this map it should be particularly pointed out if there is any intervening ground on the course of the lodes that has not been legally granted, so that proper applications may be made in due time, and not leave it until the workings have been commenced, and good discoveries made, and then this landowner, taking advantage of the neglect or oversight, demanding an unreasonable premium and dues for his land, or prohibiting us from driving an inch under it, on pain of knocking us to pieces with the powerful arm of the lord warden of the stanneries.

Among the many inconveniences that have arisen from this cause, I select one that occurred in this neighbourhood some eight or nine years ago. A silver mine called Wheal Sisters, in the parish of Calstock, was in full work, and just at the zenith of her glory. Everybody concerned with the mine, in the shape of London directors, local directors, managers, secretaries, agents, shareholders, &c., thought assuredly (if they ever thought at all) that all the land in the set belonged to the duchy of Cornwall, for no survey such as we have been speaking of had ever been made. But it so turned out, that a field under which the levels had just been driven was *freehold* ! And what was the consequence ? Why, forth came the proprietor in the person of Michael Williams,

Esq., of Scorrier House, and says, ' Stop ! ' Well,
but she did not stop then ; no, she went on faintly,
but in an expiring state, after exhausting her
resources in paying the peremptory demand of
Mr. Williams.   How much was that?   Exactly
5000*l.* !   Yes, and that gentleman was paid every
shilling of it ; and I believe not 10*l.* of silver ore
was broken in his ground afterward.   And let it be
known, this enormous sum was for the *dues* only ;
the little field is still the property of its original
owner.   This affair is well known, and is calculated
to put parties on their guard respecting their mining
rights and liabilities.   So much for the map of the
set or ground plan.

(2.) HORIZONTAL OR WORKING PLAN.

This is the miner's plan, his chart, his guide, his
right hand.   Whoever attempts to conduct the
operations of a mine without a perfect working
plan, is unfit for his office.   The very circumstance
of his supposing himself capable of doing so is a
certain proof of his ignorance.

This plan gives what surveyors call ' a bird's
eye view ' of the mine ; or let us suppose that the
ground was transparent, and by walking over every
part at the surface we could look down and
distinctly see all the workings.

A person who never saw a mine will understand
from this view that he could distinguish the course
of the levels in all their turns and windings, and,

o

as respects all the 'horizontal' drivings, he would have a *true* view of them; but these drivings are the *only* thing that he would obtain a true view of in this plan.

Keeping his position in view, he hardly requires to be told, that he can only see the brace or mouth or base of all the vertical or downright shafts, even if they were 200 fathoms deep.

As for the diagonal shafts or winzes (which are small ventilating shafts sunk on the declination of the lode from level to level), he would only see the 'underlay' of them, or the distance that they diverge from a perpendicular line. As lodes, almost if not altogether without an exception, have a dip or declination, called by miners 'underlay,' it follows that the levels are generally removed away from the vertical line, and not concealed one by another, although this is sometimes partially the case when there is a reverse or change of underlay. In addressing myself to the miner, in reply to his question or inquiry respecting the best method of constructing and keeping up a working plan, I will endeavour to explain the system I always adopt, and which I believe is the best, at least I have found it so, after thirty years' experience. Let the scale be five fathoms to an inch. Before you begin to lay down any part of the workings, draw faint lines throughout the whole length and breadth of your sheet of drawing paper at right angles, forming two-inch squares; these lines will be your cardinal points. If your lode bears east and west, the longest way of the paper will of course be appropriated for that bearing. These lines are always to remain; and as they are to be single, and fine, and the course of your levels drawn with double lines, they will not in the least confuse, especially as your levels must be distinctly and variously coloured between the double lines (which represent the breadth of the level), so that every level, with all its drifts and connections, may be distinguished in a moment or two, however numerous or complicated your levels and drivings may be. One grand advantage of these cardinal cross lines is, that every intersection forms a proper and suitable point for laying on the centre of the double-limb protractor or any other, on all occasions, so that in keeping up the plan or laying down any additional drafts, there is always a point close at hand for the protractor, without the inconvenience and risk of bringing on a north and south line, for the purpose, from a distant part of the paper. Another convenience of these lines is, that the bearing of the lode, or any part of it, may, by their help, be obtained in a few seconds; for instance, as every side of one of the squares gives 10 fathoms, when a level has been laid down, we can, by inspection, see very nearly (and by the application of a scale exactly) what it has diverged to the right or left from the main course, and if we find it to be (say) 125 fathoms east, and $19\frac{1}{2}$ fathoms north, the tables will tell us that the bearing is 8° 50' north of east; and, by the same means, we may always check or prove the truth of the plan or the construction by trigonometrical computation, and which should always be done before the plan is relied on, or pronounced perfect. This plan, proved and well kept up, becomes invaluable to the mine agent. Does he want to sink a winze in one level and rise against it in another? Everything he can wish for is before his eyes. The two corresponding points for the sink and rise, the amount of underlay, the bearing, the length of the winze and its vertical depth, are all embodied in the plan. Has the lode split, and have the workmen driven on the wrong branch? look at the plan, and compare notes with the general bearings, and the course to be adopted will be apparent. I have known a case where the plan betokened that a misdriving had taken place in a level, but the agent persisted that the driving was right in spite of the plan; however, the manager, having more confidence in *computation* than in *conceit*, was convinced by the indications of the plan that they were gone off the main branch, and ordered them immediately to 'turn house,' or cut north at right angles: this was done, and in driving two fathoms, the main lode was discovered with a large and rich course of copper ore.

(3.) Longitudinal Section.

This drawing supposes that a section of the ground has been cut away, and that a side view of the mine is exposed. If it is an east and west run, the observer is placed at the south of the mine, and taking a panoramic north view of all the excavations.

In this position he will have a perfect sight of all the vertical shafts, and a general view of the stopes, or ore ground broken away between the levels, also the dip of the courses of ore may be portrayed and distinguished, and the surface line of the country, with a perspective view of the buildings and machinery, may be seen or exhibited fairly by this section. But the levels, diagonal shafts, cross-cuts, and winzes, will have a false or imperfect appearance here. For instance, the levels will *appear* to be perfectly straight, however serpentine or crooked their course may be. The diagonal shafts and winzes will appear to be perpendicular, because their dip is in the line of the inspector's eye ; and as an *end* view will be taken of the north and south cross-cuts, the extent of these drivings will not be seen.

The only real benefit of this section to the miner is, that it may be so contrived as to show the dip, or inclination, or declination, of the bunches or courses of ore, and this circumstance he may turn greatly to his advantage in working the mine. For example, suppose in driving the 50 fathom level, going east, we cut into a course of ore, and it lasted 25 fathoms in length ; let these two points of the 'coming in' and 'going out' of the course of ore be correctly marked in this 50 fathom level of this section.

In the 60 fathom level, or next level below, the same course of ore was cut 4 fathoms farther west than it was in the 50, and the course of ore at this level proved to be 28 fathoms long. Let these points also be marked on the section, and as there is a general regularity in the dip of ores, the agent is now in possession of a clue, whereby he may form a reasonable judgment at what place the course of ore will come in at the 70 fathom level, or levels still deeper, and also at what point it will fail in driving east, hence he will be better qualified for setting tribute with the help of this section than if he had no such guide. The longitudinal sketches that are usually shown in mines, with a pell-mell blotch of the stopes, and, as we have shown, the false view of the levels, diagonals, and winzes, are useless to the miner, and deceptive to the stranger.

(4.) TRANSVERSE SECTION.

Here the view is taken at one end of the workings. Suppose again the drivings to be east and west, and the dip of the lode northerly, the observer is placed at the west end, with his face easterly. Now, for the first time, he will have a fair view of the declination of the shafts and winzes that have been sunk on the course of the lode, and thereby he will see all the dip and variations of the lode from the surface to the bottom of the mine. Here he will see the northing and southing made by the cross-cuts, and if a vertical shaft is in sinking to take the lode at a certain depth, the point of intersection will be apparent to his view. Respecting the levels driven on the course of the lode, he will only see their western end. If there has been no diagonal shaft, but the mine has been worked by a downright sump or engine-shaft, this section will exhibit a regular and correct view of all the drifts or cross-cuts, from the shaft to the lode, and from this data, or the extreme ends of the cross-cuts, the declination of the lode will be conspicuous. The transverse view of the surface line will finish all that can be fairly seen by this drawing.

OBSERVATIONS.

After such a detail we think there will be no occasion for ' summing up,' or repeating to the inquisitive stranger, or adventurer, what may be seen, and what may not be seen, on each and every drawing. To the practical man, or with him, we may converse of the best and readiest means of making these drawings. Let us suppose the horizontal or working plan to be drawn and executed, and proved in a correct and masterly manner, and all the vertical shafts truly *dropped* or measured. We are then in possession of every thing necessary for drawing the two sections without going out of the office; for by parallels, or a drawing board and slides, all the shafts, winzes, &c., may be transferred from the plan to the paper prepared for the sections, with despatch and accuracy. True, we may have recourse to the dialling book for the position, length, height, and depth of the stopes and sinks; and if a perspective drawing of that part of the set where the buildings are placed should be required, a sketch must be made for that purpose.

To the learner we would observe, that, if he is about to survey a mine and draw a working plan,

let him lay down his shallow adit, or the upper
levels, first, and the others in succession; because,
wherever any *crossings* take place, or one level or
draft passes immediately under another, the upper
level must be entire or unbroken, and the under
level will not be shown, as a matter of course, being necessarily obscured or concealed by that part
of the workings that passes immediately above it.
One method of proving his work as he proceeds is
as follows:—Suppose he has surveyed the adit
level, and there are four winzes communicating
with the 10 fathom level, and he has taken the
bearing, and depression, and length of those winzes,
and plotted or laid down this level and the true
base of those winzes on his working plan. He then
proceeds to survey the 10 fathom level, making
good every thing as he proceeds; and of course
when he arrives at the foot of those winzes which
he surveyed in the adit, he minutely enters in his
dialling book the mark at their foot, where he took
his diagonal observation and measurement. Then
in laying down his 10 fathom level, if all his work
has been well done, the points in those winzes will
exactly correspond· with his survey in the 10 fathom level and on the plan, and this desirable
check he may and should pursue throughout the
whole survey. It is too common in these cases, in
order to avoid the time and labour in surveying
the winzes, to ' let them take their chance,' by
merely entering their ' brace ' in one level and
' foot ' in another, and let the truth of their respective bearings and underlay depend on the

horizontal survey of the levels. This practice is reprehensible, and should never be tolerated. But with all this precaution, we advise, by all means, that every part of the plan be proved by trigonometrical computation, and the surveys by fore and back diallings. Let us suppose we have surveyed a level by double diallings. How shall we ascertain if there is a perfect correspondence? We have introduced a problem on this subject, and it is plain that the final two sums of the traverse will demonstrate either the agreement or the difference. This being done, and the underground work proved correct, we proceed to construct or draw the level on the plan, and it is most desirable that we should know if this part of the work has been well executed; and as we have computed the workings, we are furnished with a ready and certain test. Suppose we found, by computation, that the level gave, from beginning to end, 184 fathoms 3 feet of southing, and 34 fathoms 4 feet of westing. Now, applying these numbers to the plan, we shall, by the convenient help of the cardinal lines and instruments, presently prove if the latitude and longitude between the start and terminus of the level on the plan make good these lines. Lastly, I would recommend that the instruments for drawing and keeping up the working plan should be a 6 or 7 inch circular protractor, on the best principle, with double limb and vernier scale for reading off the angle, so that there may be no *guessing*, or judging by the eye, merely, for the fractional part of the degree ; also, a parallel ruler of the best kind. I prefer those rulers that travel on rollers, both for expedition and accuracy, but I admit it requires some practice to use them well. There is an advantage in those rulers, in that they have an ivory edge and a graduated scale, so that the lengths may be pointed off at the same time that the line is drawn, without using a compass or dividers; and these two instruments are all that are required for the drawing department. The parallel ruler should be a foot long, divided into thirty feet to an inch; so that any line within the extent of 360 feet can be pointed off at once.

MISCELLANEOUS.

THE following articles are extracts from some parts of my public correspondence during the controversy alluded to at the beginning of this Supplement, and I think will be found useful and satisfactory to many of my mining friends.

LEVELLING.

CURVATURE OF THE EARTH.

In cutting long leats or watercourses for mining purposes, it is necessary that allowance should be made for the ' curvature of the earth.'

RULE.

Consider the radius of the earth and the length of the levelling, in a right line, as the two sides of a right-angled triangle ; add the sum of the squares of those sides together, and extract their square root for the hypothenuse. The difference between the length of the radius and the hypothenuse will be the ' curvature.'

NOTE.

It will be found that the curvature on one mile will be 8 inches, and on two miles 2 feet 8 inches ; and although this rule will not answer for very long distances, yet it is sufficiently accurate for the first

5 or 6 miles, and consequently adapted for mining
purposes.

The diameter of the earth is 7963 miles, and the
half of it, or radius (say) 3831 miles, which it will
be convenient to reduce into inches, after it has
been squared, and, of course, the line of levelling
also.

There is reason to believe that failure has taken
place in bringing home long leats, where the *fall*
has been limited from want of proper attention to
this subject.

N.B. An easy and expeditious rule of approxima-
tion is to square the length of the line of levelling
in miles and divide by 8. The answer will appear
in inches.

STANDARD OF COPPER ORE.

Since my arrival in this city I have been requested
to furnish a plain definition of the term ' standard,'
as applied to copper ore; for I am told there still
exists among persons not conversant with practical
mining much mystery respecting its real meaning.

The word ' standard,' divested of its disguise, as
applied to mining, simply means ' the present value
of a ton of fine copper,' and to be understood as to
its practical effect, it must be associated with its two
near kinsmen, ' price ' and ' produce.' Standing separately, they may be thus defined; viz. ' standard,'
the value of a ton of copper; ' produce,' the number
of tons of copper in 100 tons of ore; ' price,'
the value of a ton of copper ore. Now it will be

seen that any two of these terms being given, the third may be found by proportion.

EXAMPLE.

Given standard, 116*l.*, and produce 8¾, the 'price' is required.

$$\begin{array}{ccccccc} & & & & \pounds & s. & d. \\ \text{As } 116 & : & 8\tfrac{3}{4} & :: & 100=10 & 3 & 0 \\ \end{array}$$

Deduct returning charge   2   15   0 per ton.

Answer   7   8   0 per ton.

£   *s.*   *d.*

Given 'price,' 10   3   0 Produce 8¾, require the 'standard.'

As 10   3   0   :   100   ::   8¾=Ans. 116.

£   *s.*   *d.*

Given 'standard,' 116, 'price,' 10   3   0 require the 'produce.'

£   *s.*   *d.*

As 10   3   0   :   100   ::   116=Ans. 8¾.

These operations mutually prove the rule, and I suppose will render the meaning and effect of the 'standard' clear to all parties.

GEOLOGY.

UNDER this head we shall confine ourselves to those mysterious phenomena in mines called heaves, slides, cross-courses, faults, up-throws, and by whatever names miners or geologists may call that sudden dislocation of lodes, or separation of the vein, by the intersection of another vein crossing it, either obliquely or at right angles, and sometimes removing or severing (apparently) one part of the lode from the other, and carrying it away a great many fathoms. When a lode has been thus separated it is called by miners 'a heave.' A 'slide' is of another character, although they are sometimes confounded together. If a lode running east, and underlying north, is intersected by a parallel vein, or another vein whose course is nearly parallel, but underlying contrary to it or south, and it is found at the intersection that the lode has been severed by the *flookany* vein, it is called a 'slide;' and the technical expressions are, 'cut out by a slide,' or 'thrown up,' or 'thrown down,' by a slide. The doctrine of 'heaves' and 'slides' is of great consequence to the miner; so that when one of those cross veins comes in, and 'the lode is lost,' as it is expressed, he may know in what direction to drive to find it again.

There are indications known to practical men whereby they can generally give a fair guess whether it is a ' left hand ' or a ' right hand ' heave ; but sad experience tells us that they are sometimes mistaken in this important matter, and long and wrong drivings of discovery have been made on these occasions, at a great loss of time and money. Therefore we conclude that in this, as in all of nature's operations, although there is a general rule, yet there are exceptions, and the usual guides of the bending or inclining of the last few fathoms of the lode, towards the severed part, with the *scroll* and particles of ore, scattered in that direction, &c. sometimes prove fallacious.

Now we are not going to step out of our place to presume to advise the miner respecting the best *indications*—we submit to his judgment ; but I think I shall obtain his vote, when I recommend that if a lode has been ' hove,' and has not been found by driving a few fathoms right and left on the cross-course, that the most advisable plan is to ascertain the true bearing of the lode, and set to drive on this bearing, ' through the country,' at the point where the separation took place.

I am borne out in giving this opinion, from the well-known facts that, although the course of a lode has been altered by a ' heave,' yet there is generally an inclining to recover the original course ; but especially as sometimes it is so disordered or ' squeezed up ' close to the cross-course, that it cannot be identified ; but by driving out a few fathoms through the country, on the course of the lode, and then 'turning house,' or driving at right angles, we shall be sure to find the lode, and being in settled country, to know it when found.

Of course we recommend this proceeding only in those vexatious cases where our efforts have been baffled in seeking for the lost lode by driving on the cross-course. We might bring forward some strong 'tried cases' that would preponderate in favour of this 'dernier ressort;' but we will only select the well-known case in point at Holmbush Copper-mine, in this neighbourhood. Nearly three years ago, in driving west they found the lode was 'hove' by a large cross-course; and after driving some distance north and south on the cross-course, and being strongly prepossessed in favour of the doctrine of 'right-hand heaves,' they discontinued driving south, and concentrated their force in exploring north in search of the 'lost lode;' and after a fruitless labour in driving some 50 or 60 fathoms on the cross-course, and west on some scattered branches, they remembered that they had cut a small vein in driving south about 6 or 7 fathoms south of the point of dislocation; and in driving a few fathoms on this disowned and rejected branch, they found it to expand into a large and fine course of copper ore, and to be the very lode they had expended so much time, labour, and money in searching for in the wrong direction.

But we would now turn from the miner to the geologist: O what strange, contradictory, and outrageous things have been published respecting the 'formation of lodes and mineral veins,' and the causes of 'heaves' and dislocations!

How amusing to the practical miner it is to read 'The Transactions of the Royal Geological Society,' and to see how wild, erroneous, and contradictory their *notions* are on those subjects! There is cer-. tainly one point in which they all seem to be unanimous, and that is, in 'rejecting the truth!' One grave professor will tell us, the lodes are 'secondary formations,' and have been subsequently filled from the surface; another says, the ore has been thrown up into them by 'volcanic action;' another, that it has been drawn into them sideways by 'electricity,' having been 'held in solution' in the adjoining rock; but not one of them appears 'to be valiant for the truth,' by declaring that the whole is a contemporaneous creation! No, they will leave the 'truth' for unsophisticated men to deal with, while they seem to take pleasure only in endeavouring to reconcile impossibilities.

I doubt not but a vast majority of my practical brethren are on my side in favour of the doctrine of 'contemporaneous formation.' As for the term 'heave,' 'slide,' 'throw,' 'separated,' or any other word betokening subsequent disturbance, it is nothing more than a name. These phenomena are the wise and admirable order of creation for the benefit of man. Miners know well (however ignorant geologists may be of the fact) how highly beneficial to mining operations this grand arrangement is; how advantageous are cross-courses in a granite or any other hard country, where they can

P

drive ten fathoms in about the same time and cost
that they could drive one fathom through the rock!
Another incalculable advantage is, that these cross-
courses, being generally composed of a stiff clay,
impervious to water (called 'flookan' by the miner),
effectually dam back the stream; and on some
hollow or porous lodes, were it not for these cross-
courses, the influx or drainage would be so extensive,
that it would be impossible to work the mine without an immoderate expense of machinery.

As to these 'heaves' being secondary things,
the doctrine is fraught with absurdity and impossi-
bility. We find 'heaves' of several fathoms, and
all the 'country' or rock in its vicinity, without a
single fracture, and not the slightest sign that ever
a hairbreadth movement or agitation had taken
place since the six days of creation! but what
must have been the mighty wreck and crash, if
some supernatural power had caused our ponderous
globe to shift some 10, 20, or, as in some cases, 80
fathoms, one part from another? Moreover, those
cross-courses, which are said by those 'secondary
men' to be either the cause or proofs of those dislocations, we often find them serpentine or zigzag
in their course, and that they are not interminable,
but often fail, and discontinue, when a change of
strata or country takes place.

How can we reconcile this with the idea of dislocation? Surely if the cross-course hove the lode,
before we receive the doctrine we must be assured
that the cross-course extended throughout the
globe, without interruption, and straight as a line!

In writing on this subject, during the controversy alluded to, I called the attention of the public to a notable circumstance that speaks loudly on the subject, and that can hardly fail to come under the notice of every observant man—I mean the miniature display of all the phenomena of lodes, crosscourses, heaves, slides, splits, faults, and everything else of the kind that may be seen in some species of stone. O what 'faithful witnesses' are those little representations where everything is exactly portrayed in perfect and exact harmony with the stupendous and magnificent order of the formations to be found in a mine of 200 fathoms!

If we look at the top of one of those stones, we are reminded of the working or horizontal plan of a mine, and the fractured end truly represents the transverse section. How deeply rooted must be the prejudices of men who will not receive this corroborating evidence!

I have before me the 'Mining Journal' of February 4, 1843, wherein my arguments on this subject were followed up by a practical man, as follows :—

To the Editor.

'Sir,

'After reading Mr. Budge's statement of facts inserted in the Supplement to the "Mining Journal" of Saturday, 28th inst., I recollect, many years ago, being at a mine called Seal Hole, situated in the parish of Saint Agnes, in Cornwall, now called

Polbcrou Consols, seeing there a flat blue killas or clay-slate stone, about four inches wide, five or six inches long, and half an inch thick; on the surface of which we could distinctly trace several rich tin lodes in miniature, so very similar to the disposition of the great metallic lodes, that it might be thought an exact representation of what you might expect to see in the plan and section of a mine. The little representatives of lodes, the largest of them perhaps not much larger than a stout pack-thread, and others less, all intersected by a little diminutive cross-course, composed of what the miners term spar, which heaved the different veins as is done in a large champion lode. From the above and other alike corroborative evidence, I am led to believe, also, that the lodes, and the country about them, were formed at one and the same time.

'S. P.'

If anything more is wanting to prove that those 'heaves' are *originals*, we may bring forward the fact well known to miners, that oftentimes a cross-course intersecting several lodes, the 'heaves' are found to differ materially in their distance, and some not 'heaved' at all, and there are known to be both right and left-hand heaves on the same cross-course.

I shall close this subject with the extract of a letter I inserted in the 'Mining Journal,' December 3, 1842 :—

To the Editor.

'Sir,

'At last I observe, with pleasure, that one of our practical miners has had the courage to come forward and openly throw down the gauntlet against the host of modern geologists. This champion for the truth is Mr. Thomas Deakin, of the Blaenavon Mines, whose letter appeared in last week's Journal. I hope his efforts will be seconded by abler hands than mine; for surely it is high time that the pernicious march of geological error should meet with a check from some quarter.

'Permit us to inquire what benefit mining has received from all the writings, lectures, societies, premiums, researches, and labours of our large body of theoretical geologists? If I am wrong, please to set me right; but I declare, I know not a single instance where any good has emanated from their exertions, to the value of a swabbing-stick! All the progress made in the discovery and working of mines has been without their help — the ancient methods of detecting or identifying a metallic lode by shodes, gossans, mineral waters, gases, &c., have received no improvement from them, although we are persuaded that a fine field for art and science is still open here; for as nature always works by general laws, we believe that if all the indications which attend a rich metallic lode could be detected, that mining would not be so much a speculation as it is at present. But what can be the cause that such a large body of talented men, devoted to the

subject, ambitious to excel, and " with all appli-
ances and means to boot," should be thus noto-
riously useless and unprofitable ? Your Blaenavon
correspondent has revealed the secret. Now, Mr.
Editor, allow me to observe that *theology* will never
disgrace the columns of the " Mining Journal ;" and if
we consult the best judges, I think they will admit
that it should ever be the basis of geology. This,
Sir, is the grand cause why the efforts of our geological societies have utterly failed—they have set
themselves against the *truth* — they have rejected
the inspired history of the creation of the world;
hence their writings and sayings are replete with
error, inconsistency, and contradiction.

' Let them begin again ; cancel what they have
written, and lay their foundation on the sublime
accounts so minutely given us in the Scriptures.
Then let them follow nature in all her grand and
stupendous subterranean operations, and they will
discover a world of harmonious wonders, and will
bring to light, to the admiration and benefit of
mankind, the cause and effect of the magnificent
order of every part of creation that is allowed to
fall under the inspection of man.

' I shall be borne out in stating my firm conviction that no sceptic ever made a good geologist ;
and whatever those men may think of themselves
who dare to write in contradiction to the Word
which the Creator has graciously condescended to
bestow on His creatures, they are no better than
practical atheists in the judgment of all men " who
believe and know the truth," and their writings are

calculated to inflict a serious injury on society. See, Sir, how these talented infidels try all they can to sap the foundation of the Christian's faith!—their first, second, and third *formations.* One thing *produced* by another. Coal *formed* of vegetable matter! and lately they have discovered that slate is a *marine production!* Oh, Sir, if you are not one of that school, let me entreat you to point your powerful pen against the doctrines of this geological demon, or, at least, give full introduction to those who feel themselves imperatively called on to come forward in defence of the truth.

'JOHN BUDGE.'

EXPLANATION OF TERMS USED IN THE FORE-GOING WORK.

Acclivity, the rising steepness of a hill.

Aliquot, such a part of a number or quantity as is contained in it so many times without a remainder.

Angle, the space between two lines which cross each other.

Base, the bottom, the foundation.

Bisect, to divide into two parts.

Complement, so much of an angle as is wanting to make a right angle.

Construction, the contriving such lines and figures as will show the truth of a problem.

Corollary, conclusion drawn from antecedent demonstration.

Cosecant, the secant of the complement of an arch to 90 degrees.

Cosine, the right sine of an arch which makes up another arch of 90 degrees.

Cotangent, the tangent of an arch which is the complement of another arch to 90 degrees.

Declination, the act of bending.

Declivity, the steepness of a hill downwards, gradual descent.

Definition, a clear and short description.

Degree, the 360th part of the arc of a circle.

Demonstrate, to prove with certainty.

Denomination, a name given to any thing.

Depression, the act of pressing down.

Desideratum, a desirable improvement in a science yet unattained.

Diagonal, a line drawn from one angle to another.

Diagram, a scheme drawn for the explanation of any thing.

Elevation, a raising or lifting up.

Geometry, the science of extension.

Horizon, the line that bounds the visible from the invisible part of the earth.

Horizontal, level with the horizon.

Hypothenuse, the longest side of a right-angled triangle.

Inaccessible, not to be approached.

Q

Intersect, to cut each other.

Intervene, to come between.

Junction, the act of joining together.

Mathematics, that science which contemplates whatever is capable of being numbered or measured.

Oblique, not perpendicular, not parallel.

Perpendicular, a line that stands upon or crosses another at right angles.

Problem, that which proposes something to be done. .

Protractor, an instrument to lay down or measure angles.

Quadrant, one fourth part of a circle containing 90 degrees.

Radius, a right line drawn from the centre to the circumference of a circle, half the diameter.

Right Angle, an angle containing 90 degrees, made by the touch of two lines perpendicular to each other.

Scale, a mathematical instrument containing lines divided into equal or unequal parts.

Scheme, a plan, a mathematical diagram.

Secant, a right line drawn from the centre of a circle that meets the tangent.

Segment, a part cut off.

Sine, a right line drawn from one end of an arc perpendicular upon the diameter.

Summary, compendium, abridgment.

Tabular, set down in tables.

Theorem, a position laid down as an acknowledged truth.

Tangent, a right line drawn without a circle perpendicular to some radius.

Triangle, a figure having three sides and angles.

Trigonometry, the science of resolving triangles.

Trisection, division into three parts.

Vertical, perpendicular to the horizon.

THE END.

▲

LONDON
PRINTED BY SPOTTISWOODE AND CO.
NEW-STREET SQUARE